Precalculus

Mehdi Rahmani-Andebili

Precalculus

Practice Problems, Methods, and Solutions

Second Edition

 Springer

Mehdi Rahmani-Andebili
Electrical Engineering
University of Alabama
Tuscaloosa, AL, USA

ISBN 978-3-031-49363-8   ISBN 978-3-031-49364-5   (eBook)
https://doi.org/10.1007/978-3-031-49364-5

This Springer imprint is published by the registered company Springer Nature Switzerland AG
The registered company address is: Gewerbestrasse 11, 6330 Cham, Switzerland

Paper in this product is recyclable.

Preface

Calculus is one of the most important courses of many majors, including engineering and science, and even some non-engineering majors like economics and business, which is taught in three successive courses at universities and colleges worldwide. Moreover, in many universities and colleges, a precalculus course is mandatory for under-prepared students as the prerequisite course of Calculus 1.

Unfortunately, some students do not have a solid background and knowledge in math and calculus when they start their education in universities or colleges. This issue prevents them from learning calculus-based courses, such as physics and engineering courses. Sometimes, the problem escalates, so they give up and leave the university. Based on my real professorship experience, students do not have a serious issue comprehending physics and engineering courses. In fact, it is the lack of enough knowledge of calculus that hinder them from understanding those courses.

Therefore, a series of calculus textbooks, covering Precalculus, Calculus 1, Calculus 2, and Calculus 3, have been prepared to help students succeed in their major. This book, *Precalculus: Practice Problems, Methods, and Solutions, Second Edition*, is the second edition of the book of *Precalculus: Practice Problems, Methods, and Solutions*, which was published in 2021. In the new version of the book, for almost each problem of the book, a new exercise has been added after the corresponding problem. Each exercise has been provided with a final answer (but without full solution) so that the student is encouraged to practice by himself/herself until he/she achieves the correct solution.

The subjects of the calculus series books are as follows:

Precalculus: Practice Problems, Methods, and Solution

- *Real Number Systems, Exponents and Radicals, and Absolute Values and Inequalities*
- *Systems of Equations*
- *Quadratic Equations*
- *Functions, Algebra of Functions, and Inverse Functions*
- *Factorization of Polynomials*
- *Trigonometric and Inverse Trigonometric Functions*
- *Arithmetic and Geometric Sequences*

Calculus 1: Practice Problems, Methods, and Solution

- *Characteristics of Functions*
- *Trigonometric Equations and Identities*
- *Limits and Continuities*
- *Derivatives and Their Applications*
- *Definite and Indefinite Integrals*

Calculus 2: Practice Problems, Methods, and Solution

- *Applications of Integration*
- *Sequences and Series and Their Applications*
- *Polar Coordinate System*
- *Complex Numbers*

Calculus 3: Practice Problems, Methods, and Solution

- *Linear Algebra and Analytical Geometry*
- *Lines, Surfaces, and Vector Functions in Three-Dimensional Coordinate System*
- *Multivariable Functions*
- *Double Integrals and Their Applications*
- *Triple Integrals and Their Applications*
- *Line Integrals and Their Applications*

The textbooks include basic and advanced calculus problems with very detailed problem solutions. They can be used as practicing study guides by students and as supplementary teaching sources by instructors. Since the problems have very detailed solutions, the textbooks are helpful for under-prepared students. In addition, they are beneficial for knowledgeable students because they include advanced problems.

In preparing the problems and solutions, care has been taken to use methods typically found in the primary instructor-recommended textbooks. By considering this key point, the textbooks are in the direction of instructors' lectures, and the instructors will not see any untaught and unusual problem solutions in their students' answer sheets.

To help students study in the most efficient way, the problems have been categorized into nine different levels. In this regard, for each problem, a difficulty level (easy, normal, or hard) and a calculation amount (small, normal, or large) have been assigned. Moreover, problems have been ordered in each chapter from the easiest problem with the smallest calculations to the most difficult problems with the largest ones. Therefore, students are suggested to start studying the textbooks from the easiest problems and continue practicing until they reach the normal and then the hardest ones. This classification can also help instructors choose their desirable problems to conduct a quiz or a test. Moreover, the classification of computation amount can help students manage their time during future exams, and instructors assign appropriate problems based on the exam duration.

Tuscaloosa, AL, USA   Mehdi Rahmani-Andebili

The Other Works Published by the Author

The author has already published the books and textbooks below with Springer Nature.

Calculus 3: Practice Problems, Methods, and Solutions, *Springer Nature*, 2023.

Calculus 2: Practice Problems, Methods, and Solutions, *Springer Nature*, 2023.

Calculus 1 (2nd Ed.): Practice Problems, Methods, and Solutions, *Springer Nature*, 2023.

Planning and Operation of Electric Vehicles in Smart Grid, *Springer Nature*, 2023.

Applications of Artificial Intelligence in Planning and Operation of Smart Grid, *Springer Nature*, 2022.

AC Electric Machines: Practice Problems, Methods, and Solutions, *Springer Nature*, 2022.

DC Electric Machines, Electromechanical Energy Conversion Principles, and Magnetic Circuit Analysis: Practice Problems, Methods, and Solutions, *Springer Nature*, 2022.

Differential Equations: Practice Problems, Methods, and Solutions, *Springer Nature*, 2022.

Feedback Control Systems Analysis and Design: Practice Problems, Methods, and Solutions, *Springer Nature*, 2022.

Power System Analysis: Practice Problems, Methods, and Solutions, *Springer Nature*, 2022.

Advanced Electrical Circuit Analysis: Practice Problems, Methods, and Solutions, *Springer Nature*, 2022.

Design, Control, and Operation of Microgrids in Smart Grids, *Springer Nature*, 2021.

Applications of Fuzzy Logic in Planning and Operation of Smart Grids, *Springer Nature*, 2021.

Operation of Smart Homes, *Springer Nature*, 2021.

AC Electrical Circuit Analysis: Practice Problems, Methods, and Solutions, *Springer Nature*, 2021.

Calculus: Practice Problems, Methods, and Solutions, *Springer Nature*, 2021.

Precalculus: Practice Problems, Methods, and Solutions, *Springer Nature*, 2021.

DC Electrical Circuit Analysis: Practice Problems, Methods, and Solutions, *Springer Nature*, 2020.

Planning and Operation of Plug-in Electric Vehicles: Technical, Geographical, and Social Aspects, *Springer Nature*, 2019.

Contents

Problems: Real Number Systems, Exponents and Radicals, and Absolute Values and Inequalities

1

Abstract

In this chapter, the basic and advanced problems of real number systems, exponents, radicals, absolute values, and inequalities are presented. To help students study the chapter in the most efficient way, the problems are categorized into different levels based on their difficulty (easy, normal, and hard) and calculation amounts (small, normal, and large). Moreover, the problems are ordered from the easiest, with the smallest computations, to the most difficult, with the largest calculations.

1.1 Real Number Systems

1.1. Which one of the numbers below exists [1]?

Difficulty level ● Easy ○ Normal ○ Hard
Calculation amount ● Small ○ Normal ○ Large

1) The minimum integer number smaller than -1.
2) The minimum irrational number larger than -1.
3) The maximum integer number smaller than -1.
4) The maximum rational number smaller than -1.

1.2. As we know, $\mathbb{R}$ is the set of real numbers, $\mathbb{Z}$ is the set of integer numbers, and N is the set of natural numbers. Which one of the choices is correct?

Difficulty level ● Easy ○ Normal ○ Hard
Calculation amount ● Small ○ Normal ○ Large

1) $N \subset \mathbb{Z} \subset \mathbb{R}$
2) $\mathbb{R} \subset \mathbb{Z} \subset N$
3) $\mathbb{R} \subset N \subset \mathbb{Z}$
4) $\mathbb{Z} \subset \mathbb{R} \subset N$

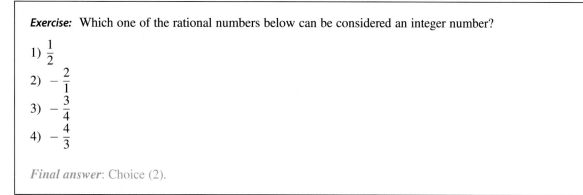

Exercise: Which one of the rational numbers below can be considered an integer number?

1) $\frac{1}{2}$

2) $-\frac{2}{1}$

3) $-\frac{3}{4}$

4) $-\frac{4}{3}$

Final answer: Choice (2).

> **Exercise:** Which one of the choices below is not an irrational number?
> 1) e
> 2) π
> 3) $0.\overline{3}$
> 4) $\sqrt{5}$
>
> **Final answer:** Choice (3).

1.3. If $\dfrac{a}{b} = \dfrac{c}{d}$, then which one of the choices is correct?

Difficulty level　　　● Easy　○ Normal　○ Hard
Calculation amount　● Small　○ Normal　○ Large

1) $\dfrac{b}{c} = \dfrac{a}{b}$

2) $\dfrac{b}{a} \neq \dfrac{d}{c}$

3) $\dfrac{a}{c} \neq \dfrac{b}{d}$

4) $\dfrac{b}{a} = \dfrac{d}{c}$

1.4. Which one of the choices has the greatest absolute value?

Difficulty level　　　○ Easy　● Normal　○ Hard
Calculation amount　● Small　○ Normal　○ Large

1) $-1 + \sqrt{2} - \sqrt{3}$
2) $-1 - \sqrt{2} + \sqrt{3}$
3) $-1 - \sqrt{2} - \sqrt{3}$
4) $1 + \sqrt{2} - \sqrt{3}$

> **Exercise:** Which one of the choices has the greatest value?
> 1) $-1 + \sqrt{2} - \sqrt{3}$
> 2) $-1 - \sqrt{2} + \sqrt{3}$
> 3) $-1 - \sqrt{2} - \sqrt{3}$
> 4) $1 + \sqrt{2} - \sqrt{3}$
>
> **Final answer:** Choice (4).

1.5. Which one of the choices below might be a rational number if $\sqrt{\alpha}$ and β^2 are rational and irrational numbers, respectively?

Difficulty level　　　○ Easy　○ Normal　● Hard
Calculation amount　● Small　○ Normal　○ Large

1) $\alpha + \beta^4$
2) $\alpha + \beta$
3) $\dfrac{\beta}{1 + \sqrt{\alpha}}$
4) $\alpha^2 + \beta$

1.2   Exponents and Radicals

1.6. Calculate the value of x^2 if $x = \sqrt[3]{2\sqrt{2}}$.

Difficulty level   ● Easy   ○ Normal   ○ Hard
Calculation amount   ● Small ○ Normal   ○ Large

1) $\sqrt{2}$
2) $\sqrt[3]{2}$
3) $\sqrt[3]{4}$
4) 2

Exercise: Simplify the term $\sqrt[3]{3\sqrt[3]{3}}$.

1) $3^{\frac{4}{9}}$

2) $3^{\frac{1}{9}}$

3) $3^{\frac{1}{3}}$

4) $3^{\frac{7}{9}}$

Final answer: Choice (1).

1.7. Calculate the value of $\left(\frac{1}{\sqrt{2}} - 1\right)^{-1}$.

Difficulty level   ● Easy   ○ Normal   ○ Hard
Calculation amount   ● Small ○ Normal   ○ Large

1) $-1 + \sqrt{2}$
2) $-2 + \sqrt{2}$
3) $-\left(1 + \sqrt{2}\right)$
4) $-\left(2 + \sqrt{2}\right)$

Exercise: Calculate the value of $\left(\sqrt{2} + 1\right)\left(\sqrt{2} - 1\right)$.

1) 1
2) -1
3) $\sqrt{2} - 2$
4) $2\sqrt{2}$

Final answer: Choice (1).

1.8. Simplify and calculate the final answer of $\sqrt[3]{(-x)^3} + \sqrt{x^2} + \sqrt{(-2)^2}$ if $x > 0$.

Difficulty level   ○ Easy   ● Normal   ○ Hard
Calculation amount   ● Small ○ Normal   ○ Large

1) $-2x - 2$
2) -2
3) $2x + 2$
4) 2

Exercise: Calculate the final answer of $\sqrt[3]{(-x)^3} + \sqrt{x^2}$ if $x < 0$.

1) $2x$
2) $-2x$
3) 0
4) $-x^3 + x^2$

Final answer: Choice (2).

1.9. Calculate the final answer of $\left(-\sqrt[10]{3^6}\right)^{\frac{5}{3}}$.

Difficulty level   ○ Easy   ● Normal   ○ Hard
Calculation amount   ● Small   ○ Normal   ○ Large

1) -9
2) 9
3) -3
4) 3

Exercise: Calculate the value of $\left(-\sqrt[9]{8^3}\right)^{\frac{1}{3}}$.

1) $\sqrt[3]{2}$
2) $\sqrt{2}$
2) $-\sqrt{2}$
3) $-\sqrt[3]{2}$

Final answer: Choice (4).

1.10. Calculate the value of $2\sqrt[3]{x^3} + \sqrt[4]{x^4}$ if $x < 0$.

Difficulty level   ○ Easy   ● Normal   ○ Hard
Calculation amount   ● Small   ○ Normal   ○ Large

1) $3x$
2) x
3) $-x$
4) $-3x$

1.11. In the equation below, determine the value of x.

$$3^{x-1} = \left(\frac{1}{81}\right)^{-8}$$

Difficulty level   ○ Easy   ● Normal   ○ Hard
Calculation amount   ● Small   ○ Normal   ○ Large

1) 24
2) 27
3) 31
4) 33

Exercise: In the following equation, calculate the value of x.

$$5^{x-1} = \left(\frac{1}{25}\right)^{-2}$$

1) 5
2) 4
3) 3
4) 2

Final answer: Choice (1).

1.12. Calculate the value of $2^{2^k} + 1$ for $k = 3$.

Difficulty level   ○ Easy   ● Normal   ○ Hard
Calculation amount   ● Small   ○ Normal   ○ Large

1) 33
2) 65
3) 129
4) 257

Exercise: Calculate the value of 2^{2^2}.

1) 16
2) 4
3) 8
4) 32

Final answer: Choice (1).

1.13. Solve the equation below.

$$\left(\frac{3}{7}\right)^{3x-7} = \left(\frac{7}{3}\right)^{7x-3}$$

Difficulty level   ○ Easy   ● Normal   ○ Hard
Calculation amount   ● Small   ○ Normal   ○ Large

1) -1
2) 1
3) $\frac{3}{7}$
4) $\frac{7}{3}$

Exercise: Solve the following equation.

$$\left(\frac{a}{b}\right)^{-x-1} = \left(\frac{b}{a}\right)^{-x+1}$$

1) 1
2) 2
3) 3
4) 0

Final answer: Choice (4).

1.14. Calculate the value of the following term.

$$\left(\frac{8}{25}\right)^{-3} \times (0.8)^4 \times 0.2$$

Difficulty level   ○ Easy   ● Normal   ○ Hard
Calculation amount   ○ Small   ● Normal   ○ Large

1) $\frac{2}{5}$

2) 2

3) $\frac{5}{2}$

4) 5

Exercise: Calculate the value of the term below.

$$\left(\frac{8}{25}\right)^{2} \times (0.8)^{-2}$$

1) $\frac{25}{4}$

2) 1

3) $\frac{4}{25}$

4) $\frac{4}{5}$

Final answer: Choice (3).

1.15. Calculate the value of the term below.

$$\frac{25}{90} \times \left(\frac{3}{2}\right)^{5} \times (0.75)^{-3}$$

Difficulty level   ○ Easy   ● Normal   ○ Hard
Calculation amount   ○ Small   ● Normal   ○ Large

1) $\dfrac{5}{2}$

2) $\dfrac{10}{3}$

3) 5

4) $\dfrac{15}{2}$

1.16. Calculate the value of the term below.

$$16 \times \left(\frac{\sqrt{2}}{2}\right)^{6} \times (0.5)^{-6} \times 8^{-\frac{4}{3}}$$

Difficulty level   ○ Easy   ● Normal   ○ Hard

Calculation amount   ○ Small   ● Normal   ○ Large

1) $\dfrac{1}{8}$

2) 2

3) 8

4) 4

1.17. Calculate the value of the term below.

$$\frac{3 \times (45)^{6}}{(15)^{6} \times 3^{7}}$$

Difficulty level   ○ Easy   ● Normal   ○ Hard

Calculation amount   ○ Small   ● Normal   ○ Large

1) 1

2) 3

3) 5

4) 15

Exercise: Calculate the value of the term below.

$$\frac{2 \times (10)^{5}}{(5)^{6} \times 2^{5}}$$

1) $\dfrac{2}{5}$

2) $\dfrac{5}{2}$

3) 1

4) 10

Final answer: Choice (1).

1.18. Calculate the value of x in the following equation.

$$\sqrt[x]{2} = \left(\left(\left((16)^{\frac{1}{3}}\right)^{\frac{1}{2}}\right)^{\frac{1}{2}}\right)^{\frac{1}{2}}$$

Difficulty level   ○ Easy   ● Normal   ○ Hard

Calculation amount   ○ Small   ● Normal   ○ Large

1) 5
2) 6
3) 9
4) 12

Exercise: Solve the equation below.

$$\sqrt[x]{3} = \left(\left(\left((27)^{\frac{1}{3}} \right)^{\frac{1}{2}} \right)^{\frac{1}{2}} \right)^{\frac{1}{2}}$$

1) 1
2) 2
3) 4
4) 8

Final answer: Choice (4).

1.19. Calculate the value of $(x + x^{-1})^{\frac{1}{3}}$ if $x = 1 - \sqrt{2}$.

Difficulty level   ○ Easy   ● Normal   ○ Hard
Calculation amount   ○ Small   ● Normal   ○ Large

1) $-\sqrt{2}$
2) -1
3) $\sqrt[3]{2}$
4) 1

Exercise: Calculate the value of the term below.

$$\left(1 - \sqrt{3} + \left(1 - \sqrt{2} \right)^{-1} \right)^{2}$$

1) $-5 + 2\sqrt{6}$
2) $5 - 2\sqrt{6}$
3) $5 + 2\sqrt{6}$
4) $-5 - 2\sqrt{6}$

Final answer: Choice (3).

1.20. Simplify and calculate the final answer of the term below.

$$\frac{1}{\sqrt{4} + \sqrt{11}} + \frac{1}{\sqrt{11} + \sqrt{18}} + \frac{1}{\sqrt{18} + \sqrt{25}}$$

Difficulty level   ○ Easy   ● Normal   ○ Hard
Calculation amount   ○ Small   ● Normal   ○ Large

1) $\dfrac{2}{7}$

2) $\dfrac{3}{7}$

3) $\dfrac{1}{3}$

4) $\dfrac{2}{3}$

1.21. Determine the relation between A and B if:

$$A = x^{\frac{t+1}{t}}, B = x^{\frac{1}{t+1}}, t \neq 0, -1$$

Difficulty level   ○ Easy   ● Normal   ○ Hard
Calculation amount   ○ Small   ○ Normal   ● Large

1) $A^{\frac{t}{t+1}} = B^{\frac{t+1}{t}}$

2) $A^{\frac{t}{t+1}} = B^{t+1}$

3) $A^{\frac{1}{t+1}} = B^{t+1}$

4) $A^{t+1} = B^{\frac{1}{t+1}}$

1.22. Assume that a and b have different signs and $a < b$. Then, which one of the following statements is true?

Difficulty level   ○ Easy   ○ Normal   ● Hard
Calculation amount   ● Small   ○ Normal   ○ Large

1) $a^2 < b^2$

2) $a^3 < b^3$

3) $b^2 < a^3$

4) $b^3 < a^3$

Exercise: Assume that $a < b$ and $|a| > |b|$. Then, which one of the following statements is true?
1) $a^3 > b^3$
2) $a^2 > b^2$
3) $a^3 < b^3$
4) The choices of (2) and (3) are correct.

Final answer: Choice (4).

1.23. Determine the final answer of $\sqrt{4 - 2\sqrt{2}} \times \sqrt[4]{6 + 4\sqrt{2}}$.

Difficulty level   ○ Easy   ○ Normal   ● Hard
Calculation amount   ○ Small   ● Normal   ○ Large

1) $\sqrt{2}$

2) 2

3) $2\sqrt{2}$

4) 4

Exercise: Calculate the value of $\sqrt{6 - 2\sqrt{2}} \times \sqrt{6 + 2\sqrt{2}}$.

1) $\sqrt{7}$
2) $2\sqrt{7}$
3) 6
4) $2\sqrt{2}$

Final answer: Choice (2).

1.24. Calculate the value of $\sqrt[4]{7 - 4\sqrt{3}}\sqrt{2 + \sqrt{3}}$.

Difficulty level   ○ Easy   ○ Normal   ● Hard
Calculation amount   ○ Small   ● Normal   ○ Large

1) $\frac{1}{2}$
2) 1
3) $\frac{3}{2}$
4) 2

1.25. Which one of the following numbers is greatest?

Difficulty level   ○ Easy   ○ Normal   ● Hard
Calculation amount   ○ Small   ● Normal   ○ Large

1) $1 + \sqrt[2]{2}$
2) $1 + \sqrt[6]{6}$
3) $1 + \sqrt[4]{4}$
4) $1 + \sqrt[3]{3}$

Exercise: Which one of the numbers below is the smallest?

1) $\sqrt[2]{5}$
2) $\sqrt[6]{6}$
3) $\sqrt[4]{4}$
4) $\sqrt[3]{3}$

Final answer: Choice (2).

1.26. Determine the final answer of the term below.

$$\frac{4^{0.75}}{1 + \sqrt{2} + \sqrt{3}} + 9^{0.25}$$

Difficulty level   ○ Easy   ○ Normal   ● Hard
Calculation amount   ○ Small   ○ Normal   ● Large

1) $\sqrt{2} - 1$
2) 1
3) $\sqrt{2}$
4) $1 + \sqrt{2}$

1.3   Absolute Values and Inequalities

1.27. If $|-x + 1| < 2$, determine the range of x.

Difficulty level   ● Easy   ○ Normal   ○ Hard
Calculation amount   ● Small   ○ Normal   ○ Large

1) $-3 < x < -1$
2) $-3 < x < 1$
3) $-1 < x < 3$
4) $1 < x < 3$

Exercise: Determine the range of x if $|x - 1| \leq 1$.

1) $0 < x < 2$
2) $0 \leq x \leq 2$
3) $-2 < x < 0$
4) $-2 \leq x \leq 0$

Final answer: Choice (2).

1.28. What is the solution of $-1 \leq 3x - 2 \leq 1$?

Difficulty level   ● Easy   ○ Normal   ○ Hard
Calculation amount   ● Small   ○ Normal   ○ Large

1) $\frac{1}{3} \leq x \leq 1$
2) $-1 \leq x \leq 1$
3) $-1 \leq x \leq \frac{1}{3}$
4) $-2 \leq x \leq 1$

Exercise: Solve the inequality equation of $-1 \leq 2x + 1 \leq 1$

Difficulty level   ● Easy   ○ Normal   ○ Hard
Calculation amount   ● Small   ○ Normal   ○ Large

1) $-1 \leq x \leq 0$
2) $0 \leq x \leq 1$
3) $-2 \leq x \leq 0$
4) $0 \leq x \leq 2$

Final answer: Choice (1).

1.29. Which one of the choices shows the range of x in the equation of $|2x - 3| \leq 5$?

Difficulty level   ● Easy   ○ Normal   ○ Hard
Calculation amount   ● Small   ○ Normal   ○ Large

1) $[2, 6]$
2) $[-1, 4]$
3) $[-2, 2]$
4) $[-3, 1]$

Exercise: Which one of the choices shows the range of x in the equation of $|3x - 3| \leq 6$?

1) [3, 6]

2) [−3, 1]

3) [1, 3]

4) [−1, 3]

Final answer: Choice (4).

1.30. If the inequalities of $|x - 1| < 0.1$ and $A < 2x - 3 < B$ are equivalent, calculate the value of $A + B$.

Difficulty level   ○ Easy   ● Normal   ○ Hard

Calculation amount   ● Small   ○ Normal   ○ Large

1) −2.1

2) −2

3) −1.1

4) −1

Exercise: If the inequalities of $|x - 1| < 1$ and $A < x - 2 < B$ are equivalent, calculate the value of A.

1) 1

2) 0

3) 2

4) −2

Final answer: Choice (4).

1.31. Which one of the choices is equivalent to $|2x - 3| < x$?

Difficulty level   ○ Easy   ● Normal   ○ Hard

Calculation amount   ● Small   ○ Normal   ○ Large

1) $|x - 2| < 1$

2) $|x - 1| < 2$

3) $|x - 3| < 1$

4) $|x - 2| < 2$

Exercise: Which one of the choices is equivalent to $|3x - 2| \leq x$?

1) $|x - 1| \leq 0.25$

2) $|x - 0.75| \leq 0.25$

3) $|x - 0.5| \leq 0.5$

4) $|x - 1| \leq 0.5$

Final answer: Choice (2).

1.32. Solve the inequality equation below.

$$\frac{1}{x-1} > \frac{1}{x-3}$$

Difficulty level   ○ Easy   ● Normal   ○ Hard
Calculation amount   ● Small   ○ Normal   ○ Large
1) $x < 3$
2) $1 < x < 3$
3) $2 < x < 3$
4) $-2 < x < 3$

1.33. Solve the inequality equation of $|x - 1| \geq |x - 3|$.
Difficulty level   ○ Easy   ● Normal   ○ Hard
Calculation amount   ● Small   ○ Normal   ○ Large
1) $|x| \leq 2$
2) $|x - 2| \leq 1$
3) $x \leq 2$
4) $x \geq 2$

Exercise: Solve the inequality equation of $|x - 2| \leq |x - 4|$.
Difficulty level   ○ Easy   ● Normal   ○ Hard
Calculation amount   ● Small   ○ Normal   ○ Large
1) $x \geq 3$
2) $|x - 2| \leq 0$
3) $x \leq 2$
4) $x \leq 3$

Final answer: Choice (4).

1.34. Solve the inequality of $|x + 1| \geq |x - 1|$.
Difficulty level   ○ Easy   ● Normal   ○ Hard
Calculation amount   ● Small   ○ Normal   ○ Large
1) $x \leq 0$
2) $x \geq 0$
3) $x \geq 1$
4) $x \leq -1$

1.35. Which one of the choices below is equivalent to $-5 < x - 11 < 3$?
Difficulty level   ○ Easy   ● Normal   ○ Hard
Calculation amount   ● Small   ○ Normal   ○ Large
1) $|x + 5| < 2$
2) $|x| < 7$
3) $|x - 9| < 5$
4) $|x - 10| < 4$

Exercise: Which one of the choices below is equivalent to $1 \le x - 3 \le 5$?

1) $|x - 6| \le 2$
2) $|x - 3| \le 3$
3) $|x - 4| \le 5$
4) $|x - 4| \le 1$

Final answer: Choice (1).

1.36. Calculate the value of $|2x - 1| + |2 - x|$ if $-1 < x < 0$.

Difficulty level   ○ Easy   ● Normal   ○ Hard
Calculation amount   ○ Small   ● Normal   ○ Large

1) $-3 - 3x$
2) $3 - 3x$
3) $-3 + 3x$
4) $1 + x$

Exercise: Calculate the value of $|3x - 1| + |1 - x|$ if $-1 < x < 0$.

Difficulty level   ○ Easy   ● Normal   ○ Hard
Calculation amount   ○ Small   ● Normal   ○ Large

1) $4x$
2) $2 - 4x$
3) $4x - 2$
4) $2x$

Final answer: Choice (2).

1.37. Determine the neighborhood radius for the deleted symmetric neighborhood of $(3a - 7, a + 5) - \{3\}$.

Difficulty level   ○ Easy   ● Normal   ○ Hard
Calculation amount   ○ Small   ● Normal   ○ Large

1) 1
2) 2
3) 3
4) 4

Exercise: Calculate the neighborhood radius for the deleted symmetric neighborhood of $(-4, 4) - \{0\}$.

1) 2
2) 4
3) 8
4) 6

Final answer: Choice (2).

1.38. If $b < 0 < a$ and $|a| > |b|$, calculate the final answer of $|a + b| + |a| + |b|$.

     Difficulty level   ○ Easy   ○ Normal   ● Hard

     Calculation amount   ● Small ○ Normal ○ Large

     1) $-2b$

     2) $-2a$

     3) $2a$

     4) $2b$

1.39. Determine the number of answers for x in the equation $|x + 1| + |x - 3| = 2$.

     Difficulty level   ○ Easy   ○ Normal   ● Hard

     Calculation amount   ● Small ○ Normal ○ Large

     1) 0

     2) 1

     3) 2

     4) 3

Reference

1. Rahmani-Andebili, M. (2021). Precalculus – Practice Problems, Methods, and Solutions, Springer Nature, 2021.

Solutions to Problems: Real Number Systems, Exponents and Radicals, and Absolute Values and Inequalities

Abstract

In this chapter, the problems of the first chapter are fully solved, in detail, step-by-step, and with different methods.

2.1   Real Number Systems

2.1. Choice (1) is not correct because the set of integer numbers is not left-bounded [1].
Choice (2) is not correct because between any pair of real numbers, infinite irrational numbers exist.
Choice (3) is correct because -2 is the maximum integer number smaller than -1.
Choice (4) is not correct because between any pair of real numbers, infinite rational numbers exist.
Choice (3) is the answer.

2.2. Figure 2.1 shows the diagram of set of real numbers, irrational numbers, rational numbers, integer numbers, whole numbers, and natural numbers. Choice (1) is the answer.

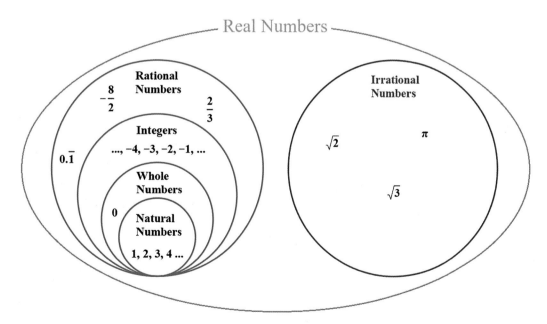

Figure 2.1 The diagram of problem 2.2

2.3. We know that:

$$\alpha = \beta \Rightarrow \frac{1}{\alpha} = \frac{1}{\beta}$$

Therefore:

$$\frac{a}{b} = \frac{c}{d} \Rightarrow \frac{b}{a} = \frac{d}{c}$$

Choice (4) is the answer.

2.4. As we know, $\sqrt{3} = 1.73$ and $\sqrt{2} = 1.41$. Therefore:

Choice (1):

$$-1 + \sqrt{2} - \sqrt{3} < 0 \Rightarrow \left| -1 + \sqrt{2} - \sqrt{3} \right| = -\left(-1 + \sqrt{2} - \sqrt{3} \right) \approx 1.31$$

Choice (2):

$$-1 - \sqrt{2} + \sqrt{3} < 0 \Rightarrow \left| -1 - \sqrt{2} + \sqrt{3} \right| = -\left(-1 - \sqrt{2} + \sqrt{3} \right) \approx 0.68$$

Choice (3):

$$-1 - \sqrt{2} - \sqrt{3} < 0 \Rightarrow \left| -1 - \sqrt{2} - \sqrt{3} \right| = -\left(-1 - \sqrt{2} - \sqrt{3} \right) \approx 4.14$$

Choice (4):

$$1 + \sqrt{2} - \sqrt{3} > 0 \Rightarrow \left| 1 + \sqrt{2} - \sqrt{3} \right| = 1 + \sqrt{2} - \sqrt{3} \approx 0.68$$

Choice (3) is the answer.

2.5. To solve the problem, we need to recall the following points:

Point 1: A rational number will remain a rational number even if it is squared.
Point 2: An irrational number might be changed to a rational number if it is squared (e.g., $\beta = \sqrt[4]{2} \Rightarrow \beta^4 = 2$).
Point 3: An irrational number will remain an irrational number if its square root is taken.
Point 4: The sum of two rational numbers is always a rational number.
Point 5: The sum of a rational number and an irrational number is always an irrational number.
Point 6: The product of a nonzero rational number and an irrational number is always an irrational number.
Point 7: A rational number will remain a rational number if it is inversed.

Choice (1) might be a rational number based on points 1, 2, and 4.
Choice (2) is an irrational number based on points 1, 3, and 5.
Choice (3) is an irrational number based on points 3, 4, 7, and 6.
Choice (4) is an irrational number based on points 1, 3, and 5.

Choice (1) is the answer.

2.2 Exponents and Radicals

2.6. We know that:

$$\sqrt[m]{a^n} = a^{\frac{n}{m}}$$

$$(a^m)^n = a^{mn}$$

Hence:

$$x = \sqrt[3]{2\sqrt{2}} = \left(2^{\frac{3}{2}}\right)^{\frac{1}{3}} = 2^{\frac{1}{2}} \Rightarrow x^2 = \left(2^{\frac{1}{2}}\right)^2 = 2$$

Choice (4) is the answer.

2.7. We know that:

$$(a+b)(a-b) = a^2 - b^2$$

Thus:

$$\left(\frac{1}{\sqrt{2}} - 1\right)^{-1} = \frac{1}{\frac{1}{\sqrt{2}} - 1} = \frac{\sqrt{2}}{1 - \sqrt{2}} = \frac{\sqrt{2}}{1 - \sqrt{2}} \times \frac{1 + \sqrt{2}}{1 + \sqrt{2}} = \frac{\sqrt{2}(1 + \sqrt{2})}{1 - 2} = -\left(\sqrt{2} + 2\right)$$

Choice (4) is the answer.

2.8. From exponent and radical rules, we know that:

$$\sqrt[n]{a^n} = a, \quad \text{if } n \text{ is an odd number}$$

$$\sqrt[n]{a^n} = |a|, \quad \text{if } n \text{ is an even number}$$

Therefore:

$$\sqrt[3]{(-x)^3} + \sqrt{x^2} + \sqrt{(-2)^2} = (-x) + |x| + |-2|$$

$$\xLongrightarrow{x > 0} -x + x + 2 = 2$$

Choice (4) is the answer.

2.9. From exponent and radical rules, we know that:

$$(ab)^n = a^n b^n$$

$$(a^m)^n = a^{mn}$$

Hence:

$$\left(-\sqrt[10]{3^6}\right)^{\frac{5}{3}} = \left(-1 \times 3^{\frac{6}{10}}\right)^{\frac{5}{3}} = (-1)^{\frac{5}{3}}\left(3^{\frac{6}{10}}\right)^{\frac{5}{3}} = \left((-1)^5\right)^{\frac{1}{3}}\left(3^{\frac{6}{10} \times \frac{5}{3}}\right) = (-1)^{\frac{1}{3}}(3^1) = -1 \times 3 = -3$$

Choice (3) is the answer.

2.10. From exponent and radical rules, we know that:

$$\sqrt[n]{a^n} = a, \quad \text{if } n \text{ is an odd number}$$

$$\sqrt[n]{a^n} = |a|, \quad \text{if } n \text{ is an even number}$$

Therefore:

$$2\sqrt[3]{x^3} + \sqrt[4]{x^4} = 2x + |x|$$

$$\xrightarrow{x < 0} 2x - x = x$$

Choice (2) is the answer.

2.11. We know that:

$$\left(\frac{1}{a}\right)^{-n} = a^n$$

Therefore:

$$\left(\frac{1}{81}\right)^{-8} = 81^8 = \left(3^4\right)^8 = 3^{32}$$

$$3^{x-1} = 3^{32} \Rightarrow x - 1 = 32 \Rightarrow x = 33$$

Choice (4) is the answer.

2.12. From exponent rules, we know that:

$$a^{n^m} = a^{\overbrace{n \times \ldots \times n}^{m \ times}}$$

Therefore, for $k = 3$, we can write:

$$2^{2^k} + 1 = 2^{2^3} + 1 = 2^8 + 1 = 256 + 1 = 257$$

Choice (4) is the answer.

2.13. From exponent rules, we know that:

$$a^n = \left(\frac{1}{a}\right)^{-n}$$

$$a^n = a^m \Rightarrow n = m$$

Therefore:

$$\left(\frac{3}{7}\right)^{3x-7} = \left(\frac{7}{3}\right)^{7x-3} \Rightarrow \left(\frac{3}{7}\right)^{3x-7} = \left(\frac{3}{7}\right)^{-7x+3}$$

$$\Rightarrow 3x - 7 = -7x + 3 \Rightarrow 10x = 10 \Rightarrow x = 1$$

Choice (2) is the answer.

2.14. From exponent rules, we know that:

$$a^n = \left(\frac{1}{a}\right)^{-n}$$

$$\left(\frac{a}{b}\right)^n = \frac{a^n}{b^n}$$

$$(a^m)^n = a^{mn}$$

$$\frac{a^n}{a^m} = a^{n-m}$$

The problem can be solved as follows:

$$\left(\frac{8}{25}\right)^{-3} = \left(\frac{25}{8}\right)^3 = \frac{\left(5^2\right)^3}{\left(2^3\right)^3} = \frac{5^6}{2^9}$$

$$(0.8)^4 = \left(\frac{8}{10}\right)^4 = \frac{\left(2^3\right)^4}{(2 \times 5)^4} = \frac{2^{12}}{2^4 \times 5^4} = \frac{2^8}{5^4}$$

$$0.2 = \frac{2}{10} = \frac{1}{5}$$

Therefore:

$$\left(\frac{8}{25}\right)^{-3} \times (0.8)^4 \times 0.2 = \frac{5^6}{2^9} \times \frac{2^8}{5^4} \times \frac{1}{5} = \frac{5^6}{5^5} \times \frac{2^8}{2^9} = 5 \times \frac{1}{2} = \frac{5}{2}$$

Choice (3) is the answer.

2.15. From exponent rules, we know that:

$$a^n = \left(\frac{1}{a}\right)^{-n}$$

$$\left(\frac{a}{b}\right)^n = \frac{a^n}{b^n}$$

$$(a^m)^n = a^{mn}$$

$$\frac{a^n}{a^m} = a^{n-m}$$

Therefore:

$$\frac{25}{90} \times \left(\frac{3}{2}\right)^5 \times (0.75)^{-3} = \frac{5}{18} \times \left(\frac{3}{2}\right)^5 \times \left(\frac{3}{4}\right)^{-3} = \frac{5}{2 \times 3^2} \times \frac{3^5}{2^5} \times \frac{4^3}{3^3} = \frac{5}{2 \times 3^2} \times \frac{3^5}{2^5} \times \frac{\left(2^2\right)^3}{3^3}$$

$$= 5 \times \frac{3^5}{3^2 \times 3^3} \times \frac{2^6}{2 \times 2^5} = 5 \times \frac{3^5}{3^5} \times \frac{2^6}{2^6} = 5 \times 1 \times 1 = 5$$

Choice (3) is the answer.

2.16. From exponent rules, we know that:

$$a^n = \left(\frac{1}{a}\right)^{-n}$$

$$\left(\frac{a}{b}\right)^n = \frac{a^n}{b^n}$$

$$(a^m)^n = a^{mn}$$

$$\frac{a^n}{a^m} = a^{n-m}$$

Hence:

$$16 \times \left(\frac{\sqrt{2}}{2}\right)^6 \times (0.5)^{-6} \times 8^{-\frac{4}{3}} = 2^4 \times \left(\frac{1}{2^{\frac{1}{2}}}\right)^6 \times \left(\frac{1}{2}\right)^{-6} \times (2^3)^{-\frac{4}{3}}$$

$$= 2^4 \times \frac{1}{2^3} \times 2^6 \times \frac{1}{2^4} = \frac{2^4 \times 2^6}{2^4 \times 2^3} = 2^{4+6-4-3} = 2^3 = 8$$

Choice (3) is the answer.

2.17. From exponent rules, we know that:

$$(a^m)^n = a^{mn}$$

$$(ab)^n = a^n b^n$$

Thus:

$$\frac{3 \times (45)^6}{(15)^6 \times 3^7} = \frac{3 \times (5 \times 3^2)^6}{(5 \times 3)^6 \times 3^7} = \frac{3 \times 5^6 \times 3^{12}}{5^6 \times 3^6 \times 3^7} = \frac{3^{1+12}}{3^{6+7}} = 1$$

Choice (1) is the answer.

2.18. From exponent rules, we know that:

$$\sqrt[m]{2} = 2^{\frac{1}{m}}$$

$$(a^m)^n = a^{mn}$$

$$a^n = a^m \Rightarrow n = m$$

Hence:

$$\sqrt[x]{2} = 2^{\frac{1}{x}}$$

$$\left(\left(\left((16)^{\frac{1}{3}}\right)^{\frac{1}{2}}\right)^{\frac{1}{2}}\right)^{\frac{1}{2}} = \left(\left(\left((2^4)^{\frac{1}{3}}\right)^{\frac{1}{2}}\right)^{\frac{1}{2}}\right)^{\frac{1}{2}} = 2^{4 \times \frac{1}{3} \times \frac{1}{2} \times \frac{1}{2} \times \frac{1}{2}} = 2^{\frac{1}{6}}$$

Therefore:

$$\sqrt[x]{2} = \left(\left(\left((16)^{\frac{1}{3}} \right)^{\frac{1}{2}} \right)^{\frac{1}{2}} \right)^{\frac{1}{2}} \Rightarrow 2^{\frac{1}{x}} = 2^{\frac{1}{6}} \Rightarrow x = 6$$

Choice (2) is the answer.

2.19. We know that:

$$(a + b)(a - b) = a^2 - b^2$$

$$(a^m)^n = a^{mn}$$

Based on the information given in the problem, we have:

$$x = 1 - \sqrt{2}$$

Therefore:

$$(x + x^{-1})^{\frac{1}{3}} = \left(1 - \sqrt{2} + \frac{1}{1 - \sqrt{2}} \right)^{\frac{1}{3}} = \left(1 - \sqrt{2} + \frac{1}{1 - \sqrt{2}} \times \frac{1 + \sqrt{2}}{1 + \sqrt{2}} \right)^{\frac{1}{3}} = \left(1 - \sqrt{2} + \frac{1 + \sqrt{2}}{1 - 2} \right)^{\frac{1}{3}}$$

$$= \left(1 - \sqrt{2} - 1 - \sqrt{2} \right)^{\frac{1}{3}} = \left(-2\sqrt{2} \right)^{\frac{1}{3}} = \left(\left(-\sqrt{2} \right)^3 \right)^{\frac{1}{3}} = -\sqrt{2}$$

Choice (1) is the answer.

2.20. Based on the rule of difference of two squares, we know that:

$$\left(\sqrt{a} + \sqrt{b} \right) \left(\sqrt{a} - \sqrt{b} \right) = a - b$$

To solve this problem, we need to change the denominator of each fraction to a rational number as follows:

$$\frac{1}{\sqrt{4} + \sqrt{11}} + \frac{1}{\sqrt{11} + \sqrt{18}} + \frac{1}{\sqrt{18} + \sqrt{25}}$$

$$= \frac{1}{\sqrt{4} + \sqrt{11}} \times \frac{\sqrt{4} - \sqrt{11}}{\sqrt{4} - \sqrt{11}} + \frac{1}{\sqrt{11} + \sqrt{18}} \times \frac{\sqrt{11} - \sqrt{18}}{\sqrt{11} - \sqrt{18}} + \frac{1}{\sqrt{18} + \sqrt{25}} \times \frac{\sqrt{18} - \sqrt{25}}{\sqrt{18} - \sqrt{25}}$$

$$= \frac{\sqrt{4} - \sqrt{11}}{4 - 11} + \frac{\sqrt{11} - \sqrt{18}}{11 - 18} + \frac{\sqrt{18} - \sqrt{25}}{18 - 25} = \frac{\sqrt{4} - \sqrt{11} + \sqrt{11} - \sqrt{18} + \sqrt{18} - \sqrt{25}}{-7}$$

$$= \frac{\sqrt{4} - \sqrt{25}}{-7} = \frac{2 - 5}{-7} = \frac{3}{7}$$

Choice (2) is the answer.

2.21. From the problem, we have:

$$A = x^{\frac{t+1}{t}}, B = x^{\frac{1}{t+1}}, t \neq 0, -1$$

From exponent rules, we know that:

$$(a^m)^n = a^{mn}$$

In choice (1):

$$\begin{cases} A^{\frac{t}{t+1}} = \left(x^{\frac{t+1}{t}}\right)^{\frac{t}{t+1}} = x \\ B^{\frac{t+1}{t}} = \left(x^{\frac{1}{t+1}}\right)^{\frac{t+1}{t}} = x^{\frac{1}{t}} \end{cases} \Rightarrow A^{\frac{t}{t+1}} \neq B^{\frac{t+1}{t}}$$

In choice (2):

$$\begin{cases} A^{\frac{t}{t+1}} = \left(x^{\frac{t+1}{t}}\right)^{\frac{t}{t+1}} = x \\ B^{t+1} = \left(x^{\frac{1}{t+1}}\right)^{t+1} = x \end{cases} \Rightarrow A^{\frac{t}{t+1}} = B^{t+1}$$

In choice (3):

$$\begin{cases} A^{\frac{1}{t+1}} = \left(x^{\frac{t+1}{t}}\right)^{\frac{1}{t+1}} = x^{\frac{1}{t}} \\ B^{t+1} = \left(x^{\frac{1}{t+1}}\right)^{t+1} = x \end{cases} \Rightarrow A^{\frac{1}{t+1}} \neq B^{t+1}$$

In choice (4):

$$\begin{cases} A^{t+1} = \left(x^{\frac{t+1}{t}}\right)^{t+1} = x^{\frac{(t+1)^2}{t}} \\ B^{\frac{1}{t+1}} = \left(x^{\frac{1}{t+1}}\right)^{\frac{1}{t+1}} = x^{\frac{1}{(t+1)^2}} \end{cases} \Rightarrow A^{t+1} \neq B^{\frac{1}{t+1}}$$

Choice (2) is the answer.

2.22. Based on the information given in the problem, we know that a and b have different signs and $a < b$. Therefore:

$$a < 0, b > 0$$

$$\overset{()^3}{\Rightarrow} a^3 < 0, b^3 > 0 \Rightarrow a^3 < b^3$$

Choice (2) is the answer.

2.23. We know that:

$$(a+b)(a-b) = a^2 - b^2$$

$$\sqrt[m]{a^n} = a^{\frac{n}{m}}$$

The problem can be solved as follows:

$$\sqrt{4 - 2\sqrt{2}} \times \sqrt[4]{6 + 4\sqrt{2}} = \sqrt{4 - 2\sqrt{2}} \times \sqrt[4]{2^2 + \left(\sqrt{2}\right)^2 + 4\sqrt{2}} = \sqrt{2\left(2 - \sqrt{2}\right)} \times \sqrt[4]{\left(2 + \sqrt{2}\right)^2}$$

$$= \sqrt{2\left(2 - \sqrt{2}\right)} \times \sqrt{2 + \sqrt{2}} = \sqrt{2\left(2 - \sqrt{2}\right)\left(2 + \sqrt{2}\right)} = \sqrt{2(4 - 2)} = \sqrt{4} = 2$$

Choice (2) is the answer.

2.24. We know that:

$$(a+b)(a-b) = a^2 - b^2$$

$$\sqrt[m]{a^n} = a^{\frac{n}{m}}$$

Therefore:

$$\sqrt[4]{7-4\sqrt{3}}\sqrt{2+\sqrt{3}} = \sqrt[4]{7-4\sqrt{3}}\sqrt[4]{\left(2+\sqrt{3}\right)^2}$$

$$= \sqrt[4]{\left(7-4\sqrt{3}\right)\left(2+\sqrt{3}\right)^2} = \sqrt[4]{\left(7-4\sqrt{3}\right)\left(4+4\sqrt{3}+3\right)}$$

$$= \sqrt[4]{\left(7-4\sqrt{3}\right)\left(7+4\sqrt{3}\right)} = \sqrt[4]{(7)^2 - \left(4\sqrt{3}\right)^2} = \sqrt[4]{49-48} = \sqrt[4]{1} = 1$$

Choice (2) is the answer.

2.25. From radical rules, we know that:

$$\sqrt[m]{a} = \sqrt[m \times n]{a^n}$$

Now, we should define the radicals based on a common root (greatest common divisor (GCL)), as follows:

$$GCL\{2, 6, 4, 3\} = 12$$

Therefore:

$$1 + \sqrt[2]{2} = 1 + \sqrt[2 \times 6]{2^6} = 1 + \sqrt[12]{64}$$

$$1 + \sqrt[6]{6} = 1 + \sqrt[6 \times 2]{6^2} = 1 + \sqrt[12]{36}$$

$$1 + \sqrt[4]{4} = 1 + \sqrt[4 \times 3]{4^3} = 1 + \sqrt[12]{64}$$

$$1 + \sqrt[3]{3} = 1 + \sqrt[3 \times 4]{3^4} = 1 + \sqrt[12]{81}$$

As can be seen, $1 + \sqrt[3]{3}$ is the greatest number. Choice (4) is the answer.

2.26. We know that:

$$(a+b)(a-b) = a^2 - b^2$$

$$(a^m)^n = a^{mn}$$

The problem can be solved as follows:

$$\frac{4^{0.75}}{1+\sqrt{2}+\sqrt{3}} + 9^{0.25} = \frac{\left(2^2\right)^{\frac{3}{4}}}{1+\sqrt{2}+\sqrt{3}} + \left(3^2\right)^{\frac{1}{4}} = \frac{2^{\frac{3}{2}}}{1+\sqrt{2}+\sqrt{3}} + 3^{\frac{1}{2}} = \frac{2\sqrt{2}}{1+\sqrt{2}+\sqrt{3}} + \sqrt{3}$$

$$= \frac{2\sqrt{2}}{1+\sqrt{2}+\sqrt{3}} \times \frac{1+\sqrt{2}-\sqrt{3}}{1+\sqrt{2}-\sqrt{3}} + \sqrt{3} = \frac{2\sqrt{2}\left(1+\sqrt{2}-\sqrt{3}\right)}{\left(1+\sqrt{2}\right)^2 - \left(\sqrt{3}\right)^2} + \sqrt{3} = \frac{2\sqrt{2}\left(1+\sqrt{2}-\sqrt{3}\right)}{1+2\sqrt{2}+2-3} + \sqrt{3}$$

$$= \frac{2\sqrt{2}\left(1+\sqrt{2}-\sqrt{3}\right)}{2\sqrt{2}} + \sqrt{3} = 1 + \sqrt{2} - \sqrt{3} + \sqrt{3} = 1 + \sqrt{2}$$

Choice (4) is the answer.

2.3   Absolute Values and Inequalities

2.27. We know from the features of absolute value and inequality that:

$$|x-a| < b \Leftrightarrow -b < x-a < b$$

The problem can be solved as follows:

$$|-x+1| < 2 \Rightarrow -2 < -x+1 < 2 \xrightarrow{-1} -3 < -x < 1 \xrightarrow{\times(-1)} -1 < x < 3$$

Choice (3) is the answer.

2.28. The problem can be solved as follows:

$$-1 \le 3x-2 \le 1 \xrightarrow{+2} 1 \le 3x \le 3 \xrightarrow{\times\frac{1}{3}} \frac{1}{3} \le x \le 1$$

Choice (1) is the answer.

2.29. We know from the features of absolute value and inequality that:

$$|x-a| < b \Leftrightarrow -b < x-a < b$$

The problem can be solved as follows:

$$|2x-3| \le 5 \Rightarrow -5 \le 2x-3 \le 5 \xrightarrow{+3} -2 \le 2x \le 8 \xrightarrow{\times\frac{1}{2}} -1 \le x \le 4$$

Choice (2) is the answer.

2.30. We know from the features of absolute value and inequality that:

$$|x-a| < b \Leftrightarrow -b < x-a < b$$

Based on the information given in the problem, we have:

$$|x-1| < 0.1 \Leftrightarrow A < 2x-3 < B \tag{1}$$

The problem can be solved as follows:

$$|x-1|<0.1 \Rightarrow -0.1<x-1<0.1 \xrightarrow{\times 2} -0.2<2x-2<0.2 \xrightarrow{-1} -1.2<2x-3<-0.8 \ (2) \qquad (2)$$

By comparing (1) and (2), we can conclude that:

$$A=-1.2, B=-0.8 \Rightarrow A+B=-2$$

Choice (2) is the answer.

2.31. We know from the features of absolute value and inequality that:

$$|x-a|<b \Leftrightarrow -b<x-a<b$$

The problem can be solved as follows:

$$|2x-3|<x \Rightarrow -x<2x-3<x \Rightarrow \begin{cases} 2x-3<x \Rightarrow x<3 \\ -x<2x-3 \Rightarrow 1<x \end{cases} \Rightarrow 1<x<3 \xrightarrow{-2} -1<x-2<1 \Rightarrow |x-2|<1$$

Choice (1) is the answer.

2.32. The problem can be solved as follows:

$$\frac{1}{x-1}>\frac{1}{x-3} \Rightarrow \frac{1}{x-1}-\frac{1}{x-3}>0 \Rightarrow \frac{(x-3)-(x-1)}{(x-1)(x-3)}>0 \Rightarrow \frac{-2}{(x-1)(x-3)}>0$$

$$\Rightarrow (x-1)(x-3)<0 \Rightarrow 1<x<3$$

Choice (2) is the answer.

2.33. The problem can be solved as follows:

$$|x-1| \geq |x-3| \xrightarrow{(\ )^2} (x-1)^2 \geq (x-3)^2 \Rightarrow x^2-2x+1 \geq \Rightarrow x^2-6x+9 \Rightarrow 4x \geq 8 \Rightarrow x \geq 2$$

Choice (4) is the answer.

2.34. The problem can be solved as follows:

$$|x+1| \geq |x-1| \xrightarrow{(\ )^2} (x+1)^2 \geq (x-1)^2 \Rightarrow x^2+2x+1 \geq \Rightarrow x^2-2x+1 \Rightarrow 4x \geq 0 \Rightarrow x \geq 0$$

Choice (2) is the answer.

2.35. We know from the features of absolute value and inequality that:

$$|x-a|<b \Leftrightarrow -b<x-a<b$$

The problem can be solved as follows:

$$-5<x-11<3 \xrightarrow{+11} 6<x<14 \xrightarrow{-10} -4<x-10<4 \Rightarrow |x-10|<4$$

Choice (4) is the answer.

2.36. The problem can be solved as follows:

$$-1 < x < 0 \xrightarrow{\times 2} -2 < 2x < 0 \xrightarrow{-1} -3 < 2x-1 < -1 \Rightarrow |2x-1| < 1-2x$$

$$-1 < x < 0 \xrightarrow{\times(-1)} 0 < -x < 1 \xrightarrow{+2} 2 < 2-x < 3 \Rightarrow |2-x| < 2-x$$

$$\Rightarrow |2x-1| + |2-x| = 1-2x+2-x = 3-3x$$

Choice (2) is the answer.

2.37. The neighborhood center (C) and neighborhood radius (R) for the symmetric neighborhood of $(\alpha_1, \alpha_3) - \{\alpha_2\}$ can be determined as follows:

$$C = \alpha_2 = \frac{\alpha_1 + \alpha_3}{2} \tag{1}$$

$$R = \frac{\alpha_3 - \alpha_1}{2} \tag{2}$$

Therefore, for $(3a - 7, a + 5) - \{3\}$, we can write:

$$3 = \frac{(3a-7) + (a+5)}{2} \Rightarrow 3 = \frac{4a-2}{2} \Rightarrow a = 2 \tag{3}$$

$$R = \frac{(a+5) - (3a-7)}{2} = \frac{-2a+12}{2} \tag{4}$$

Solving (3) and (4):

$$R = \frac{-2 \times 2 + 12}{2} = 4$$

Choice (4) is the answer.

2.38. The problem can be solved as follows:

$$a > 0 \Rightarrow |a| = a \tag{1}$$

$$b < 0 \Rightarrow |b| = -b \tag{2}$$

$$|a| > |b| \xrightarrow{(1),(2)} a > -b \Rightarrow a+b > 0 \Rightarrow |a+b| = a+b \tag{3}$$

Solving (1), (2), and (3):

$$|a+b| + |a| + |b| = a+b+a+(-b) = 2a$$

Choice (3) is the answer.

2.39. From the problem, we have:

$$|x + 1| + |x - 3| = 2 \tag{1}$$

We know that:

$$|A| + |B| \geq |A - B|$$

Therefore:

$$|x + 1| + |x - 3| \geq |x + 1 - (x - 3)| \Rightarrow |x + 1| + |x - 3| \geq 4 \tag{2}$$

Solving (1) and (2):

$$2 \geq 4 \Rightarrow \text{Wrong}$$

Therefore, there is no answer to the problem. Choice (1) is the answer.

Reference

1. Rahmani-Andebili, M. (2021). Precalculus – Practice Problems, Methods, and Solutions, Springer Nature, 2021.

Abstract

In this chapter, the basic and advanced problems of systems of equations are presented. To help students study the chapter in the most efficient way, the problems are categorized into different levels based on their difficulty (easy, normal, and hard) and calculation amounts (small, normal, and large). Moreover, the problems are ordered from the easiest, with the smallest computations, to the most difficult, with the largest calculations.

3.1. For what value of m, the system of equations below has a unique solution [1]?

$$\begin{cases} 2x - 3y = 5 \\ x + my = 2 \end{cases}$$

Difficulty level   ● Easy   ○ Normal   ○ Hard
Calculation amount   ● Small   ○ Normal   ○ Large

1) $m = \frac{3}{2}$
2) $m \neq -\frac{3}{2}$
3) $m = -\frac{3}{2}$
4) For any m

Exercise: For what value of a, the following system of equations has a unique solution?

$$\begin{cases} ax + 4y = 3 \\ x + ay = 4 \end{cases}$$

1) $a = 2$
2) $a \neq 2$
3) $a \neq \pm 2$
4) For any value of a

Final answer: Choice (3).

3.2. If the value of x in the system of equations below is 2.5, determine the value of $2a + b$.

$$\begin{cases} ax - y = 1 \\ bx + 2y = 3 \end{cases}$$

Difficulty level   ○ Easy   ● Normal   ○ Hard
Calculation amount   ● Small   ○ Normal   ○ Large
1) -2
2) -1
3) 1
4) 2

Exercise: Calculate the value of $a + 3b$ for $x = 1$ in the following system of equations.

$$\begin{cases} ax - 3y = 1 \\ bx + y = 3 \end{cases}$$

1) 1
2) 3
3) 5
4) 10

Final answer: Choice (4).

3.3. Calculate the value of $y^2 + y$ from the solution of the system of equations below.

$$\begin{cases} (x + 1)^2 + 3y = 8 \\ x(x + 2) - 5y = 31 \end{cases}$$

Difficulty level   ○ Easy   ● Normal   ○ Hard
Calculation amount   ● Small   ○ Normal   ○ Large
1) 12
2) 6
3) 10
4) 8

3.4. For what value of a, the matrix below has a unique solution?

$$\begin{bmatrix} a + 1 & 2 \\ -1 & a - 1 \end{bmatrix} \begin{bmatrix} x \\ y \end{bmatrix} = \begin{bmatrix} a \\ 1 \end{bmatrix}$$

Difficulty level   ○ Easy   ● Normal   ○ Hard
Calculation amount   ● Small   ○ Normal   ○ Large
1) $\{-1, 1\}$
2) $\mathbb{R} - \{0, 1\}$
3) For no value of a
4) $\mathbb{R}$

Exercise: The matrix below has a unique solution. Determine the value of m.

$$\begin{bmatrix} m & 1 \\ -1 & m+2 \end{bmatrix} \begin{bmatrix} x \\ y \end{bmatrix} = \begin{bmatrix} 2 \\ 1 \end{bmatrix}$$

1) $\{-1\}$
2) $\mathbb{R} - \{-1\}$
3) For no value of a
4) For any value of a

Final answer: Choice (2).

3.5. Calculate the value of y from the solution of the system of equations below.

$$\begin{cases} \dfrac{x}{2} - \dfrac{y}{3} = \dfrac{4}{3} \\ 2x + 3y = 14 \end{cases}$$

Difficulty level   ○ Easy   ● Normal   ○ Hard
Calculation amount   ● Small   ○ Normal   ○ Large
1) -4
2) 2
3) 3
4) 4

3.6. Solve the system of equations below.

$$\begin{cases} \dfrac{x-1}{2} = \dfrac{y-1}{3} \\ x + 2y = 11 \end{cases}$$

Difficulty level   ○ Easy   ● Normal   ○ Hard
Calculation amount   ● Small   ○ Normal   ○ Large
1) $x = -3, y = -4$
2) $x = 3, y = 4$
3) $x = 3, y = -4$
4) $x = -3, y = 4$

Exercise: Solve the following system of equations.

$$\begin{cases} x + y = 2 \\ 2x + 3y = 3 \end{cases}$$

1) $x = 3, y = 1$
2) $x = -3, y = 1$
3) $x = 3, y = -1$
4) $x = -3, y = -1$

Final answer: Choice (3).

3.7. Calculate the value of $x + y$ from the solution of the system of equations below.

$$\begin{cases} x + 2y = -1 \\ 3x - y = 4 \end{cases}$$

Difficulty level   ○ Easy   ● Normal   ○ Hard
Calculation amount   ● Small   ○ Normal   ○ Large
1) -1
2) 0
3) 1
4) 2

3.8. For what value of a, the system of equations below has an infinite number of solutions?

$$\begin{cases} 2x + 3y = 4 \\ y = a(x - 2) \end{cases}$$

Difficulty level   ○ Easy   ● Normal   ○ Hard
Calculation amount   ● Small   ○ Normal   ○ Large
1) $-\dfrac{3}{2}$

2) $-\dfrac{2}{3}$

3) $\dfrac{2}{3}$

4) $\dfrac{3}{2}$

Exercise: The system of equations below has an infinite number of solutions. Determine the value of a.

$$\begin{cases} ax + y = 4 \\ x + ay = 4a \end{cases}$$

1) ± 1
2) -1
3) 1
4) $\mathbb{R}$

Final answer: Choice (1).

3.9. Calculate the value of $y - x$ from the solution of the system of equations below.

$$\begin{cases} x + 3y = -2 \\ 3x + y = -14 \end{cases}$$

Difficulty level   ○ Easy   ● Normal   ○ Hard
Calculation amount   ● Small   ○ Normal   ○ Large
1) -4
2) -3
3) 3
4) 6

Exercise: Calculate the value of $x - y$ from the solution of the system of equations below.

$$\begin{cases} x + 4y = 5 \\ 2x + 3y = 5 \end{cases}$$

1) 8
2) 0
3) -10
4) -8

Final answer: Choice (2).

3.10. Calculate the value of y from the solution of the system of equations below.

$$\begin{cases} \dfrac{x}{y} = 2 \\ x + 2y = 4 \end{cases}$$

Difficulty level   ○ Easy   ● Normal   ○ Hard
Calculation amount   ● Small   ○ Normal   ○ Large
1) 4
2) 3
3) 2
4) 1

3.11. Calculate the value of xy from the solution of the system of equations below.

$$\begin{cases} 3x + 2y = -1 \\ 2x + 3y = -4 \end{cases}$$

Difficulty level   ○ Easy   ● Normal   ○ Hard
Calculation amount   ● Small   ○ Normal   ○ Large
1) -3
2) -2
3) 2
4) 3

Exercise: Calculate the value of xy from the solution of the system of equations below.

$$\begin{cases} x + 3y = 5 \\ 2x + 2y = -4 \end{cases}$$

1) $-\dfrac{77}{4}$
2) 2
3) -2
4) $\dfrac{77}{4}$

Final answer: Choice (1).

3.12. Calculate the determinant of the matrix of A if:

$$A^{-1} = \begin{bmatrix} 2 & 3 \\ -7 & 6 \end{bmatrix}$$

Difficulty level   ○ Easy   ● Normal   ○ Hard
Calculation amount   ● Small   ○ Normal   ○ Large

1) $\dfrac{1}{33}$

2) $\dfrac{1}{9}$

3) 1

4) 33

Exercise: Calculate the determinant of the matrix of A.

$$A = \begin{bmatrix} 3 & -2 \\ -1 & 4 \end{bmatrix}$$

1) 10
2) 14
3) 12
4) 8

Final answer: Choice (1).

3.13. Calculate the matrix of X if we have:

$$AX = 2I, A = \begin{bmatrix} -2 & -1 \\ 4 & 3 \end{bmatrix}$$

Difficulty level   ○ Easy   ● Normal   ○ Hard
Calculation amount   ○ Small   ● Normal   ○ Large

1) $\begin{bmatrix} -3 & -1 \\ 4 & 2 \end{bmatrix}$

2) $\begin{bmatrix} 3 & 1 \\ 4 & -2 \end{bmatrix}$

3) $2\begin{bmatrix} -3 & -1 \\ 4 & 2 \end{bmatrix}$

4) $\dfrac{1}{2}\begin{bmatrix} 3 & 1 \\ 4 & -2 \end{bmatrix}$

Exercise: Calculate the matrix of X in the following equation.

$$\begin{bmatrix} 3 & 5 \\ 1 & 2 \end{bmatrix} X = I$$

1) $\begin{bmatrix} 2 & -5 \\ -1 & 3 \end{bmatrix}$

2) $\begin{bmatrix} 2 & 5 \\ -1 & 3 \end{bmatrix}$

3) $\begin{bmatrix} 2 & -5 \\ 1 & 3 \end{bmatrix}$

4) $\begin{bmatrix} 2 & 5 \\ 1 & 3 \end{bmatrix}$

Final answer: Choice (1).

3.14. Calculate the value of $x + y$.

$$\begin{bmatrix} 2 & -1 \\ 1 & -1 \end{bmatrix} \begin{bmatrix} x \\ y \end{bmatrix} = \begin{bmatrix} 3 \\ 1 \end{bmatrix}$$

Difficulty level   ○ Easy   ● Normal   ○ Hard
Calculation amount   ○ Small   ● Normal   ○ Large
1) 1
2) 2
3) 3
4) 4

3.15. Which one of the following matrices is invertible?
Difficulty level   ○ Easy   ● Normal   ○ Hard
Calculation amount   ○ Small   ● Normal   ○ Large

1) $\begin{bmatrix} 2 & 4 \\ 3 & 5 \end{bmatrix}$

2) $\begin{bmatrix} 2 & 4 \\ 3 & 6 \end{bmatrix}$

3) $\begin{bmatrix} 2 & 8 \\ 3 & 12 \end{bmatrix}$

4) $\begin{bmatrix} -2 & -1 \\ 6 & 3 \end{bmatrix}$

Exercise: Which one of the matrices below is not invertible?

1) $\begin{bmatrix} 1 & 4 \\ 3 & 5 \end{bmatrix}$

2) $\begin{bmatrix} 2 & 4 \\ 3 & 2 \end{bmatrix}$

3) $\begin{bmatrix} 30 & 5 \\ 3 & 20 \end{bmatrix}$

4) $\begin{bmatrix} 4 & 1 \\ 8 & 2 \end{bmatrix}$

Final answer: Choice (4).

3.16. Calculate the value of b if:

$$A = \begin{bmatrix} -1 & b \\ 0 & 1 \end{bmatrix}, \quad A^{-1} = \begin{bmatrix} -1 & 3 \\ 0 & 1 \end{bmatrix}$$

Difficulty level   ○ Easy   ● Normal   ○ Hard
Calculation amount   ○ Small   ● Normal   ○ Large
1) 0
2) 1
3) 2
4) 3

Exercise: Calculate the value of $a + b$ if:

$$\begin{bmatrix} 2 & 5 \\ 1 & 3 \end{bmatrix}^{-1} = \begin{bmatrix} a & b \\ -1 & 2 \end{bmatrix}$$

1) 2
2) −2
3) 5
4) −1

Final answer: Choice (2).

3.17. For what value of m, the system of equations below has an infinite number of solutions?

$$\begin{cases} m(x-1) = 3(x-y) \\ 4x + (m+1)y = 2 \end{cases}$$

Difficulty level   ○ Easy   ● Normal   ○ Hard
Calculation amount   ○ Small   ● Normal   ○ Large

1) 2
2) −3
3) 3
4) 5

3.18. For what value of m, the system of equations below does not have a solution?

$$\begin{cases} my + 2x = 5 \\ y + (m-1)x = 2m - 3 \end{cases}$$

Difficulty level   ○ Easy   ● Normal   ○ Hard
Calculation amount   ○ Small   ● Normal   ○ Large

1) −2
2) −1
3) 1
4) 2

Exercise: For what value of a, the system of equations below does not have a solution?

$$\begin{cases} y + 2x = 6 \\ y + (a-1)x = a \end{cases}$$

1) 1
2) 2
3) 3
4) 4

Final answer: Choice (3).

3.19. Calculate the value of $x + y$ from the solution of the following system of equations.

$$\begin{cases} \dfrac{3}{2x-1} - \dfrac{1}{y-2} = \dfrac{4}{3} \\ \dfrac{1}{2x-1} + \dfrac{5}{2-y} = 2 \end{cases}$$

Difficulty level   ○ Easy   ● Normal   ○ Hard
Calculation amount   ○ Small   ○ Normal   ● Large

1) 1
2) 2
3) 3
4) 4

3.20. How many solution(s) does the system of equations below have?

$$\frac{5x + y}{1} = \frac{3x - y}{3} = \frac{7x + y}{2}$$

Difficulty level   ○ Easy   ○ Normal   ● Hard
Calculation amount   ○ Small   ● Normal   ○ Large

1) 1
2) 2
3) No solution
4) Infinite number of solutions

Reference

1. Rahmani-Andebili, M. (2021). Precalculus – Practice Problems, Methods, and Solutions, Springer Nature, 2021.

Solutions to Problems: Systems of Equations

4

Abstract

In this chapter, the problems of the third chapter are fully solved, in detail, step-by-step, and with different methods.

4.1. The system of equations with the form presented in (1) has a unique solution (the lines intersect each other) if [1]:

$$\frac{a_1}{a_2} \neq \frac{b_1}{b_2}$$

$$\begin{cases} a_1 x + b_1 y = c_1 \\ a_2 x + b_2 y = c_2 \end{cases} \tag{1}$$

Therefore:

$$\begin{cases} 2x - 3y = 5 \\ x + my = 2 \end{cases} \Rightarrow \frac{2}{1} \neq \frac{-3}{m} \Rightarrow m \neq -\frac{3}{2}$$

Choice (2) is the answer.

4.2. The problem can be solved as follows:

$$\begin{cases} ax - y = 1 \\ bx + 2y = 3 \end{cases} \xrightarrow{x = 2.5} \begin{cases} 2.5a - y = 1 \\ 2.5b + 2y = 3 \end{cases} \overset{\times 2}{\Rightarrow} \begin{cases} 5a - 2y = 2 \\ 2.5b + 2y = 3 \end{cases} \overset{+}{\Rightarrow} 5a + 2.5b = 5 \overset{\times \frac{2}{5}}{\Rightarrow} 2a + b = 2$$

Choice (4) is the answer.

4.3. The problem can be solved as follows:

$$\begin{cases} (x+1)^2 + 3y = 8 \\ x(x+2) - 5y = 31 \end{cases} \Rightarrow \begin{cases} x^2 + 2x + 3y = 7 \\ x^2 + 2x - 5y = 31 \end{cases} \overset{-}{\Rightarrow} 8y = -24 \Rightarrow y = -3 \Rightarrow y^2 = 9$$

$$\Rightarrow y^2 + y = 9 + (-3) = 6$$

Choice (2) is the answer.

4.4. The system of equations with the matrix form presented in (1) has a solution (the lines intersect each other) if its determinant is nonzero as follows:

$$|A| \neq 0 \Rightarrow a_1 a_4 - a_2 a_3 \neq 0$$

$$AX = B \Rightarrow \begin{bmatrix} a_1 & a_2 \\ a_3 & a_4 \end{bmatrix} \begin{bmatrix} x_1 \\ x_2 \end{bmatrix} = \begin{bmatrix} b_1 \\ b_2 \end{bmatrix} \tag{1}$$

Therefore:

$$\begin{bmatrix} a+1 & 2 \\ -1 & a-1 \end{bmatrix} \begin{bmatrix} x \\ y \end{bmatrix} = \begin{bmatrix} a \\ 1 \end{bmatrix} \Rightarrow |A| = (a+1)(a-1) - (2)(-1) = a^2 - 1 + 2 \neq 0 \Rightarrow a^2 + 1 \neq 0 \Rightarrow a \in \mathbb{R}$$

Choice (4) is the answer.

4.5. The problem can be solved as follows:

$$\begin{cases} \dfrac{x}{2} - \dfrac{y}{3} = \dfrac{4}{3} \\ 2x + 3y = 14 \end{cases} \xrightarrow[\times(-3)]{\times 12} \begin{cases} 6x - 4y = 16 \\ -6x - 9y = -42 \end{cases} \xrightarrow{+} -13y = -26 \Rightarrow y = 2$$

Choice (2) is the answer.

4.6. The problem can be solved as follows:

$$\begin{cases} \dfrac{x-1}{2} = \dfrac{y-1}{3} \\ x + 2y = 11 \end{cases} \xrightarrow{\times 6} \begin{cases} 3x - 3 = 2y - 2 \\ x + 2y = 11 \end{cases} \Rightarrow \begin{cases} 3x - 2y = 1 \\ x + 2y = 11 \end{cases} \xrightarrow{+} 4x = 12 \Rightarrow x = 3$$

$$x + 2y = 11 \xrightarrow{x=3} 3 + 2y = 11 \Rightarrow y = 4$$

Choice (2) is the answer.

4.7. The problem can be solved as follows:

$$\begin{cases} x + 2y = -1 \\ 3x - y = 4 \end{cases} \Rightarrow \begin{cases} x + 2y = -1 \\ 6x - 2y = 8 \end{cases} \xrightarrow{+} 7x = 7 \Rightarrow x = 1$$

$$3x - y = 4 \xrightarrow{x=1} 3 \times 1 - y = 4 \Rightarrow y = -1$$

$$\Rightarrow x + y = 1 + (-1) = 0$$

Choice (2) is the answer.

4.8. The system of equations with the form presented in (1) has an infinite number of solutions (the lines match each other) if:

$$\frac{a_1}{a_2} = \frac{b_1}{b_2} = \frac{c_1}{c_2}$$

$$\begin{cases} a_1 x + b_1 y = c_1 \\ a_2 x + b_2 y = c_2 \end{cases} \tag{1}$$

Thus:

$$\begin{cases} 2x + 3y = 4 \\ y = a(x-2) \end{cases} \Rightarrow \begin{cases} 2x + 3y = 4 \\ ax - y = 2a \end{cases} \Rightarrow \frac{2}{a} = \frac{3}{-1} = \frac{4}{2a} \Rightarrow a = -\frac{2}{3}$$

Choice (2) is the answer.

4.9. The problem can be solved as follows:

$$\begin{cases} x + 3y = -2 \\ 3x + y = -14 \end{cases} \xrightarrow[\Rightarrow]{\times(-3)} \begin{cases} -3x - 9y = 6 \\ 3x + y = -14 \end{cases} \overset{+}{\Rightarrow} -8y = -8 \Rightarrow y = 1$$

$$3x + y = -14 \xrightarrow{y=1} 3x + 1 = -14 \Rightarrow x = -5$$

$$\Rightarrow y - x = 1 - (-5) = 6$$

Choice (4) is the answer.

4.10. The problem can be solved as follows:

$$\begin{cases} \frac{x}{y} = 2 \\ x + 2y = 4 \end{cases} \Rightarrow \begin{cases} x - 2y = 0 \\ x + 2y = 4 \end{cases} \xrightarrow[\Rightarrow]{\times(-1)} \begin{cases} -x + 2y = 0 \\ x + 2y = 4 \end{cases} \overset{+}{\Rightarrow} 4y = 4 \Rightarrow y = 1$$

Choice (4) is the answer.

4.11. The problem can be solved as follows:

$$\begin{cases} 3x + 2y = -1 \\ 2x + 3y = -4 \end{cases} \xrightarrow[\overset{\times 3}{\Rightarrow}]{\times(-2)} \begin{cases} -6x - 4y = 2 \\ 6x + 9y = -12 \end{cases} \overset{+}{\Rightarrow} 5y = -10 \Rightarrow y = -2$$

$$2x + 3y = -4 \xrightarrow{y=-2} 2x + 3 \times (-2) = -4 \Rightarrow 2x = 2 \Rightarrow x = 1$$

$$\Rightarrow xy = 1 \times (-2) = -2$$

Choice (2) is the answer.

4.12. The determinant of a matrix is determined as follows:

$$A = \begin{bmatrix} a_1 & a_2 \\ a_3 & a_4 \end{bmatrix} \Rightarrow |A| = \begin{vmatrix} a_1 & a_2 \\ a_3 & a_4 \end{vmatrix} = a_1 a_4 - a_2 a_3$$

The relation below exists between the determinants of a matrix and its inverse matrix.

$$|A^{-1}| = \frac{1}{|A|}$$

Therefore:

$$A^{-1} = \begin{bmatrix} 2 & 3 \\ -7 & 6 \end{bmatrix} \Rightarrow |A^{-1}| = \begin{vmatrix} 2 & 3 \\ -7 & 6 \end{vmatrix} = 2 \times 6 - (-7) \times 3 = 12 + 21 = 33$$

$$\Rightarrow |A| = \frac{1}{|A^{-1}|} = \frac{1}{33}$$

Choice (1) is the answer.

4.13. The system of equations in matrix form can be solved as follows:

$$AX = B \Rightarrow X = A^{-1}B$$

$$\begin{bmatrix} a_1 & a_2 \\ a_3 & a_4 \end{bmatrix} \begin{bmatrix} x_1 & x_2 \\ x_3 & x_4 \end{bmatrix} = \begin{bmatrix} b_1 & b_2 \\ b_3 & b_4 \end{bmatrix} \Rightarrow \begin{bmatrix} x_1 & x_2 \\ x_3 & x_4 \end{bmatrix} = \begin{bmatrix} a_1 & a_2 \\ a_3 & a_4 \end{bmatrix}^{-1} \begin{bmatrix} b_1 & b_2 \\ b_3 & b_4 \end{bmatrix}$$

$$\Rightarrow \begin{bmatrix} x_1 & x_2 \\ x_3 & x_4 \end{bmatrix} = \frac{1}{a_1 a_4 - a_2 a_3} \begin{bmatrix} a_4 & -a_2 \\ -a_3 & a_1 \end{bmatrix} \begin{bmatrix} b_1 & b_2 \\ b_3 & b_4 \end{bmatrix}$$

Therefore:

$$AX = 2I \Rightarrow \begin{bmatrix} -2 & -1 \\ 4 & 3 \end{bmatrix} \begin{bmatrix} x_1 & x_2 \\ x_3 & x_4 \end{bmatrix} = 2 \begin{bmatrix} 1 & 0 \\ 0 & 1 \end{bmatrix} = \begin{bmatrix} 2 & 0 \\ 0 & 2 \end{bmatrix}$$

$$\Rightarrow \begin{bmatrix} x_1 & x_2 \\ x_3 & x_4 \end{bmatrix} = \begin{bmatrix} -2 & -1 \\ 4 & 3 \end{bmatrix}^{-1} \begin{bmatrix} 2 & 0 \\ 0 & 2 \end{bmatrix} = \frac{1}{(-2)(3) - (4)(-1)} \begin{bmatrix} 3 & 1 \\ -4 & -2 \end{bmatrix} \begin{bmatrix} 2 & 0 \\ 0 & 2 \end{bmatrix} = -\frac{1}{2} \begin{bmatrix} 6 & 2 \\ -8 & -4 \end{bmatrix} = \begin{bmatrix} -3 & -1 \\ 4 & 2 \end{bmatrix}$$

Choice (1) is the answer.

4.14. The system of equations in matrix form can be solved as follows:

$$AX = B \Rightarrow X = A^{-1}B$$

$$\begin{bmatrix} a_1 & a_2 \\ a_3 & a_4 \end{bmatrix} \begin{bmatrix} x_1 \\ x_2 \end{bmatrix} = \begin{bmatrix} b_1 \\ b_2 \end{bmatrix} \Rightarrow \begin{bmatrix} x_1 \\ x_2 \end{bmatrix} = \begin{bmatrix} a_1 & a_2 \\ a_3 & a_4 \end{bmatrix}^{-1} \begin{bmatrix} b_1 \\ b_2 \end{bmatrix}$$

$$\Rightarrow \begin{bmatrix} x_1 \\ x_2 \end{bmatrix} = \frac{1}{a_1 a_4 - a_2 a_3} \begin{bmatrix} a_4 & -a_2 \\ -a_3 & a_1 \end{bmatrix} \begin{bmatrix} b_1 \\ b_2 \end{bmatrix}$$

Therefore:

$$\begin{bmatrix} 2 & -1 \\ 1 & -1 \end{bmatrix}\begin{bmatrix} x \\ y \end{bmatrix} = \begin{bmatrix} 3 \\ 1 \end{bmatrix} \Rightarrow \begin{bmatrix} x \\ y \end{bmatrix} = \frac{1}{(2)(-1)-(1)(-1)}\begin{bmatrix} -1 & 1 \\ -1 & 2 \end{bmatrix}\begin{bmatrix} 3 \\ 1 \end{bmatrix} = \begin{bmatrix} 1 & -1 \\ 1 & -2 \end{bmatrix}\begin{bmatrix} 3 \\ 1 \end{bmatrix} = \begin{bmatrix} 2 \\ 1 \end{bmatrix}$$

$$\Rightarrow x = 2, y = 1 \Rightarrow x + y = 3$$

Choice (3) is the answer.

4.15. A matrix is invertible if its determinant is nonzero. In other words:

$$A = \begin{bmatrix} a_1 & a_2 \\ a_3 & a_4 \end{bmatrix} \Rightarrow |A| = \begin{vmatrix} a_1 & a_2 \\ a_3 & a_4 \end{vmatrix} \neq 0 \Rightarrow a_1 a_4 - a_2 a_3 \neq 0$$

Choice (1):

$$\begin{bmatrix} 2 & 4 \\ 3 & 5 \end{bmatrix} \Rightarrow |A| = \begin{vmatrix} 2 & 4 \\ 3 & 5 \end{vmatrix} = 2 \times 5 - 4 \times 3 = -2 \neq 0$$

Choice (2):

$$\begin{bmatrix} 2 & 4 \\ 3 & 6 \end{bmatrix} \Rightarrow |A| = \begin{vmatrix} 2 & 4 \\ 3 & 6 \end{vmatrix} = 2 \times 6 - 4 \times 3 = 0$$

Choice (3):

$$\begin{bmatrix} 2 & 8 \\ 3 & 12 \end{bmatrix} \Rightarrow |A| = \begin{vmatrix} 2 & 8 \\ 3 & 12 \end{vmatrix} = 2 \times 12 - 8 \times 3 = 0$$

Choice (4):

$$\begin{bmatrix} -2 & -1 \\ 6 & 3 \end{bmatrix} \Rightarrow |A| = \begin{vmatrix} -2 & -1 \\ 6 & 3 \end{vmatrix} = (-2) \times 3 - 6 \times (-1) = 0$$

Choice (1) is the answer.

4.16. The inverse of a matrix can be determined as follows:

$$A = \begin{bmatrix} a_1 & a_2 \\ a_3 & a_4 \end{bmatrix} \Rightarrow A^{-1} = \begin{bmatrix} a_1 & a_2 \\ a_3 & a_4 \end{bmatrix}^{-1} = \frac{1}{a_1 a_4 - a_2 a_3}\begin{bmatrix} a_4 & -a_2 \\ -a_3 & a_1 \end{bmatrix}$$

Therefore:

$$A = \begin{bmatrix} -1 & b \\ 0 & 1 \end{bmatrix} \Rightarrow A^{-1} = \begin{bmatrix} -1 & b \\ 0 & 1 \end{bmatrix}^{-1} = \frac{1}{(-1) \times 1 - 0 \times b}\begin{bmatrix} 1 & -b \\ 0 & -1 \end{bmatrix} \Rightarrow A^{-1} = \begin{bmatrix} -1 & b \\ 0 & 1 \end{bmatrix} \tag{2}$$

Based on the information given in the problem, we have:

$$A^{-1} = \begin{bmatrix} -1 & 3 \\ 0 & 1 \end{bmatrix} \tag{1}$$

Solving (1) and (2):

$$\begin{bmatrix} -1 & 3 \\ 0 & 1 \end{bmatrix} = \begin{bmatrix} -1 & b \\ 0 & 1 \end{bmatrix} \Rightarrow b = 3$$

Choice (4) is the answer.

4.17. The system of equations with the form presented in (1) has an infinite number of solutions (the lines match each other) if:

$$\frac{a_1}{a_2} = \frac{b_1}{b_2} = \frac{c_1}{c_2}$$

$$\begin{cases} a_1 x + b_1 y = c_1 \\ a_2 x + b_2 y = c_2 \end{cases} \tag{1}$$

Hence:

$$\begin{cases} m(x-1) = 3(x-y) \\ 4x + (m+1)y = 2 \end{cases} \Rightarrow \begin{cases} (m-3)x + 3y = m \\ 4x + (m+1)y = 2 \end{cases} \Rightarrow \frac{m-3}{4} = \frac{3}{m+1} = \frac{m}{2} \Rightarrow \begin{cases} \dfrac{3}{m+1} = \dfrac{m}{2} & (2) \\ \dfrac{m-3}{4} = \dfrac{m}{2} & (3) \end{cases}$$

$$(2) \Rightarrow \frac{3}{m+1} = \frac{m}{2} \Rightarrow m^2 + m = 6 \Rightarrow m^2 + m - 6 = 0 \Rightarrow (m+3)(m-2) = 0 \Rightarrow m = -3, 2$$

However, $m = 2$ does not satisfy (3), as is shown below:

$$(3) \Rightarrow \frac{m-3}{4} = \frac{m}{2} \Rightarrow \frac{2-3}{4} = \frac{2}{2} \Rightarrow -\frac{1}{4} \neq 1$$

Therefore, only $m = -3$ is acceptable. Choice (2) is the answer.

4.18. The system of equations with the form presented in (1) does not have a solution (the lines are parallel) if:

$$\frac{a_1}{a_2} = \frac{b_1}{b_2} \neq \frac{c_1}{c_2}$$

$$\begin{cases} a_1 x + b_1 y = c_1 \\ a_2 x + b_2 y = c_2 \end{cases} \tag{1}$$

Therefore:

$$\begin{cases} my + 2x = 5 \\ y + (m-1)x = 2m - 3 \end{cases} \Rightarrow \begin{cases} 2x + my = 5 \\ (m-1)x + y = 2m - 3 \end{cases} \Rightarrow \frac{2}{m-1} = \frac{m}{1} \neq \frac{5}{2m-3} \Rightarrow \begin{cases} \dfrac{2}{m-1} = \dfrac{m}{1} & (2) \\ \dfrac{m}{1} \neq \dfrac{5}{2m-3} & (3) \end{cases}$$

$$(2) \Rightarrow \frac{2}{m-1} = \frac{m}{1} \Rightarrow m^2 - m = 2 \Rightarrow m^2 - m - 2 = 0 \Rightarrow (m-2)(m+1) = 0 \Rightarrow m = -1, 2$$

However, $m = -1$ does not satisfy (3), as is proven below:

$$\frac{m}{1} \neq \frac{5}{2m-3} \Rightarrow \frac{-1}{1} \neq \frac{5}{2 \times (-1) - 3} \Rightarrow -1 \neq -1 \Rightarrow \text{Wrong}$$

Therefore, only $m = 2$ is acceptable. Choice (4) is the answer.

4.19. The problem can be solved as follows:

$$\begin{cases} \dfrac{3}{2x-1} - \dfrac{1}{y-2} = \dfrac{4}{3} \\ \dfrac{1}{2x-1} + \dfrac{5}{2-y} = 2 \end{cases} \xrightarrow{\times (-5)} \begin{cases} \dfrac{-15}{2x-1} + \dfrac{5}{y-2} = -\dfrac{20}{3} \\ \dfrac{1}{2x-1} + \dfrac{5}{2-y} = 2 \end{cases} \xrightarrow{+} \dfrac{-15}{2x-1} + \dfrac{1}{2x-1} = -\dfrac{20}{3} + 2$$

$$\Rightarrow \frac{-14}{2x-1} = -\frac{14}{3} \Rightarrow 2x - 1 = 3 \Rightarrow x = 2$$

$$\frac{1}{2x-1} + \frac{5}{2-y} = 2 \xrightarrow{x=2} \frac{1}{2 \times 2 - 1} + \frac{5}{2-y} = 2 \Rightarrow \frac{5}{2-y} = \frac{5}{3} \Rightarrow 2 - y = 3 \Rightarrow y = -1$$

$$\Rightarrow x + y = 2 + (-1) = 1$$

Choice (1) is the answer.

4.20. The system of equations shows the equations of two lines. Therefore, we need to check each pair of equations.

$$\frac{5x+y}{1} = \frac{3x-y}{3} = \frac{7x+y}{2} \Rightarrow \begin{cases} \dfrac{5x+y}{1} = \dfrac{3x-y}{3} \\ \dfrac{3x-y}{3} = \dfrac{7x+y}{2} \end{cases} \Rightarrow \begin{cases} 15x+3y=3x-y \\ 6x-2y=21x+3y \end{cases} \Rightarrow \begin{cases} 12x+4y=0 \\ 15x+5y=0 \end{cases}$$

$$\Rightarrow \begin{cases} 3x+y=0 \\ 3x+y=0 \end{cases}$$

Therefore, the equations of the lines are the same and the lines match each other. Hence, the system of equations has an infinite number of solutions. Choice (4) is the answer.

Reference

1. Rahmani-Andebili, M. (2021). Precalculus – Practice Problems, Methods, and Solutions, Springer Nature, 2021.

Problems: Quadratic Equations

Abstract

In this chapter, the basic and advanced problems of quadratic equations are presented. To help students study the chapter in the most efficient way, the problems are categorized into different levels based on their difficulty (easy, normal, and hard) and calculation amounts (small, normal, and large). Moreover, the problems are ordered from the easiest, with the smallest computations, to the most difficult, with the largest calculations.

5.1. Two times of a positive quantity is 9 units fewer than one-third of its squared value. Determine this quantity [1].

Difficulty level   ● Easy   ○ Normal   ○ Hard
Calculation amount ● Small   ○ Normal   ○ Large
1) 9
2) 12
3) 15
4) 18

Exercise: Calculate the value of a positive nonzero quantity, where four times of this quantity is equal to one-half of its squared value of squared value.
1) 1
2) 2
3) 3
4) 4

Final answer: Choice (2).

5.2. For what value of a, the value of the term $ax^2 + 2x + 4a$ is always positive?

Difficulty level   ○ Easy   ● Normal   ○ Hard
Calculation amount ● Small   ○ Normal   ○ Large
1) $a > 0.5$
2) $a < -0.5$
3) $0 < a < 0.5$
4) $-0.5 < a < 0.5$

Exercise: For what value of a, the value of $ax^2 + 2x + 1$ is always positive?
1) $a > 1$
2) $a > 0$
3) $0 < a < 1$
4) $-1 < a < 0$

Final answer: Choice (1).

5.3. For what value of k, the roots of the quadratic equation $2x^2 + 3x - k = 0$ are two units fewer than the roots of the quadratic equation $2x^2 - 5x + 1 = 0$.

Difficulty level   ○ Easy   ● Normal   ○ Hard
Calculation amount   ● Small   ○ Normal   ○ Large
1) 1
2) 2
3) 3
4) 4

Exercise: For what value of k, the roots of the quadratic equation $x^2 + x + k = 0$ are one unit more than the roots of the quadratic equation $x^2 + 3x + 2 = 0$.
1) 0
2) 1
3) 2
4) 3

Final answer: Choice (1).

5.4. For what value of m, the quadratic equation $2x^2 - 5x + m = 0$ has two real roots that are reciprocal to each other?
Difficulty level   ○ Easy   ● Normal   ○ Hard
Calculation amount   ● Small   ○ Normal   ○ Large
1) 1
2) 2
3) 3
4) 4

Exercise: For what value of m, the roots of the quadratic equation $3x^2 + 7x + 3m = 0$ are reciprocal to each other?
1) 1
2) 2
3) 3
4) 4

Final answer: Choice (1).

5.5. Determine the value of m so that the quadratic equation $(m + 2)x^2 + 4x + m - 1 = 0$ has two real roots.
Difficulty level   ○ Easy   ● Normal   ○ Hard
Calculation amount   ● Small   ○ Normal   ○ Large

1) $-2 \leq m \leq 1$
2) $1 \leq m \leq 2$
3) $-2 \leq m \leq 2$
4) $-3 \leq m \leq 2$

Exercise: Calculate the value of m so that the quadratic equation $mx^2 + 2x + m - 1 = 0$ has two real roots.

1) $0 \leq m \leq 1$

2) $\dfrac{1 - \sqrt{3}}{2} \leq m \leq \dfrac{1 + \sqrt{3}}{2}$

3) $\dfrac{1 - \sqrt{2}}{2} \leq m \leq \dfrac{1 + \sqrt{2}}{2}$

4) $\dfrac{1 - \sqrt{5}}{2} \leq m \leq \dfrac{1 + \sqrt{5}}{2}$

Final answer: Choice (4).

5.6. For what value of m, the inequality $m^3 x^2 + mx + \frac{1}{m} < 0$ is always true?

Difficulty level   ○ Easy   ● Normal   ○ Hard
Calculation amount   ● Small   ○ Normal   ○ Large

1) $|m| < 3$
2) $m < 0$
3) $m > 0$
4) $m \in \mathbb{R}$

Exercise: For what value of m, the inequality $mx^2 + mx + \frac{1}{m}$ is always positive?

1) $-2 \leq m \leq 2$
2) $0 \leq m \leq 2$
3) $0 < m < 2$
4) $m \in \mathbb{R}$

Final answer: Choice (3).

5.7. Which one of the quadratic equations below has the root values that are 9 times the roots of the quadratic equation $x^2 + x - 3 = 0$?

Difficulty level   ○ Easy   ● Normal   ○ Hard
Calculation amount   ● Small   ○ Normal   ○ Large

1) $x^2 + 9x - 243 = 0$
2) $x^2 + 9x - 27 = 0$
3) $x^2 + 18x - 243 = 0$
4) $x^2 + 18x - 27 = 0$

Exercise: Which one of the following quadratic equations has the root values that are one-third of the roots of the quadratic equation $x^2 + 2x - 1 = 0$?

1) $\frac{1}{9}x^2 + \frac{2}{3}x - 1 = 0$

2) $x^2 + \frac{2}{3}x - \frac{1}{9} = 0$

3) $\frac{1}{9}x^2 + \frac{2}{3}x + 1 = 0$

4) $x^2 + \frac{2}{3}x - \frac{1}{3} = 0$

Final answer: Choice (2).

5.8. The sum of the roots of the quadratic equation $ax^2 + bx + c = 0$ is equal to the product of the reciprocal of the roots. What relation exists between the parameters?

Difficulty level   ○ Easy   ● Normal   ○ Hard
Calculation amount   ● Small   ○ Normal   ○ Large

1) $a^2 + bc = 0$
2) $a^2 - bc = 0$
3) $b^2 - ac = 0$
4) $b^2 + ac = 0$

Exercise: The sum of the roots of the quadratic equation $ax^2 + bx + c = 0$ is equal to the product of its roots. What relation exists between the parameters?

Difficulty level   ○ Easy   ● Normal   ○ Hard
Calculation amount   ● Small   ○ Normal   ○ Large

1) $a^2 + c = 0$
2) $a^2 - c = 0$
3) $b - c = 0$
4) $b + c = 0$

Final answer: Choice (4).

5.9. For what value of a, the quadratic equation $2x^2 + ax + a - \frac{3}{2} = 0$ has two distinct roots?

Difficulty level   ○ Easy   ● Normal   ○ Hard
Calculation amount   ● Small   ○ Normal   ○ Large

1) $a < 2 \; or \; a > 6$
2) $a < 3 \; or \; a > 4$
3) $2 < a < 6$
4) $3 < a < 4$

Exercise: For what value of a, the quadratic equation $2x^2 + \sqrt{8}x + a = 0$ has two distinct roots?

1) $a > 1$
2) $a < 1$
3) $a \geq 1$
4) $a \leq 1$

Final answer: Choice (2).

5.10. Determine the range of m so that $4x^2 - 2mx + 4m^2 \geq 0$.

Difficulty level   ○ Easy   ● Normal   ○ Hard
Calculation amount   ● Small   ○ Normal   ○ Large
1) $m \in \mathbb{R}$
2) $m \in \{ \ \}$
3) $|m| \leq 2$
4) $|m| \geq 2$

5.11. In the quadratic equation $2x^2 + (2k - 1)x - k = 0$, for what value of k, the sum of the reciprocal of the roots is equal to $\frac{7}{3}$?

Difficulty level   ○ Easy   ● Normal   ○ Hard
Calculation amount   ○ Small   ● Normal   ○ Large
1) -4
2) -3
3) 3
4) 4

Exercise: Calculate the value of k so that the sum of the reciprocal of the roots of $kx^2 + (k + 1)x + 1 = 0$ is equal to 2.
1) -1
2) -2
3) -3
4) -4

Final answer: Choice (3).

5.12. Determine the quadratic equation if its roots are $2 + \sqrt{4-a}$ and $2 - \sqrt{4-a}$.

Difficulty level   ○ Easy   ● Normal   ○ Hard
Calculation amount   ○ Small   ● Normal   ○ Large
1) $x^2 - 4x + a = 0$
2) $x^2 + ax - 4 = 0$
3) $x^2 + 4x - a = 0$
4) $x^2 - ax + 4 = 0$

Exercise: Determine the quadratic equation if its roots are $1 + \sqrt{4-a}$ and $1 - \sqrt{4-a}$.

1) $x^2 + (a - 3)x - 2 = 0$
2) $x^2 - (a - 3)x - 2 = 0$
3) $x^2 + 2x + a - 3 = 0$
4) $x^2 - 2x + a - 3 = 0$

Final answer: Choice (4).

5.13. For what value of m, the sum of the squares of the roots of $2x^2 - 5x + m = 0$ is $\dfrac{37}{4}$?

Difficulty level   ○ Easy   ● Normal   ○ Hard
Calculation amount   ○ Small   ● Normal   ○ Large
1) -2
2) -3
3) 2
4) 3

Exercise: For what value of m, the sum of the squares of the roots of $x^2 - \sqrt{2}x + m = 0$ is 1?

1) 0.5
2) 1
3) 2.5
4) 1.5

Final answer: Choice (1).

5.14. If x_1 and x_2 are the roots of the quadratic equation $mnx^2 + n^2x + m^2 = 0$, determine the value of $x_1^2 x_2 + x_1 x_2^2$.

Difficulty level   ○ Easy   ● Normal   ○ Hard
Calculation amount   ○ Small   ● Normal   ○ Large

1) -1
2) 1
3) $\dfrac{m + n}{mn}$
4) mn

Exercise: Calculate the value of $x_1^2 x_2 + x_1 x_2^2$, if x_1 and x_2 are the roots of the quadratic equation $x^2 + 4x + 3 = 0$,

1) -12
2) 12
3) 4
4) -4

Final answer: Choice (1).

5.15. For what value of m, the quadratic equation $(m + 1)x^2 + m(m^2 - 9)x - 2 = 0$ has two real and symmetric (equal in magnitude, but different in sign) roots.

Difficulty level   ○ Easy   ● Normal   ○ Hard
Calculation amount   ○ Small   ● Normal   ○ Large

1) -1
2) -3
3) 3
4) 9

5.16. What is the minimum value of $x^2 - x + 2$?

Difficulty level   ○ Easy   ● Normal   ○ Hard
Calculation amount   ○ Small   ● Normal   ○ Large

1) $\dfrac{1}{4}$

2) $\dfrac{3}{4}$

3) $\dfrac{5}{4}$

4) $\dfrac{7}{4}$

Exercise: Calculate the minimum value of $x^2 - 2x + 1$.
1) 0
2) 1
3) 2
4) 3

Final answer: Choice (1).

5.17. Determine the value of k if the relation of $x_1^2 + x_2^2 = 12$ exists for the roots of the quadratic equation $x^2 - 2kx - 2 = 0$.

Difficulty level   ○ Easy   ● Normal   ○ Hard
Calculation amount   ○ Small   ● Normal   ○ Large

1) $\pm\sqrt{2}$
2) $\pm\sqrt{3}$
3) ± 2
4) ± 3

Exercise: Calculate the value of k if $x_1^2 + x_2^2 = 5$, where x_1 and x_2 are the roots of the quadratic equation $x^2 + x - k = 0$.
1) 1
2) 2
3) 3
4) 4

Final answer: Choice (2).

5.18. The difference of the roots of the quadratic equation $2x^2 - 3x + m = 0$ is $\frac{5}{2}$. Determine the value of m.

Difficulty level   ○ Easy   ● Normal   ○ Hard
Calculation amount   ○ Small   ● Normal   ○ Large

1) -2
2) -1
3) 1
4) 2

Exercise: Determine the roots of the quadratic equation $2x^2 - 2x + a = 0$, if the difference of the roots is 2.
1) $-1.5, -0.5$
2) $1.5, 0.5$
3) $-1.5, 0.5$
4) $1.5, -0.5$

Final answer: Choice (4).

5.19. If the roots of the quadratic equation $x^2 + ax - a^2 = 0$ are x_1 and x_2, determine the quadratic equation when its roots are $x_1 + 1$ and $x_2 + 1$.

Difficulty level   ○ Easy   ○ Normal   ● Hard
Calculation amount   ● Small   ○ Normal   ○ Large

1) $x^2 + (a - 2)x + a^2 + a - 1 = 0$
2) $x^2 + (a + 2)x - a^2 - a - 1 = 0$
3) $x^2 + (a - 2)x - a^2 - a + 1 = 0$
4) $x^2 + (a + 2)x + a^2 + a - 1 = 0$

Exercise: If the roots of the quadratic equation $x^2 - 2x - 3 = 0$ are x_1 and x_2, determine the quadratic equation when its roots are $x_1 - 1$ and $x_2 - 1$.
1) $x^2 + 4 = 0$
2) $x^2 - 2x - 5 = 0$
3) $x^2 - 4 = 0$
4) $x^2 - 2x - 1 = 0$

Final answer: Choice (3).

5.20. The sum of the squares of two successive numbers is 925. Determine the sum of these two numbers.

Difficulty level   ○ Easy   ○ Normal   ● Hard
Calculation amount   ● Small   ○ Normal   ○ Large
1) 41
2) 43
3) 45
4) 47

5.21. If x' and x'' are the roots of the quadratic equation $x^2 - 4x + 1 = 0$, determine the value of $\left| \sqrt{x'} - \sqrt{x''} \right|$.

Difficulty level   ○ Easy   ○ Normal   ● Hard
Calculation amount   ○ Small   ● Normal   ○ Large
1) $\sqrt{2}$
2) $\sqrt{3}$
3) 2
4) 3

Exercise: Calculate the value of $\left| \sqrt{x'} - \sqrt{x''} \right|$, if x' and x'' are the roots of the quadratic equation $x^2 - 5x + 1 = 0$.
1) 1
2) $\sqrt{2}$
3) $\sqrt{3}$
4) 2

Final answer: Choice (3).

5.22. If the quadratic equations $x^2 - x - 2a = 0$ and $x^2 + 2x + a = 0$ have a common root, determine the common root for $a > 0$.

Difficulty level   ○ Easy   ○ Normal   ● Hard
Calculation amount   ○ Small   ● Normal   ○ Large
1) -2
2) -1
3) 1
4) 2

Exercise: If the quadratic equations $x^2 - 3x + 2a = 0$ and $x^2 + 2x + a = 0$ have a common root, determine the common root for $a > 0$.

1) $\dfrac{7}{5}$

2) -35

3) 0

4) $-\dfrac{7}{5}$

Final answer: Choice (2).

5.23. What is the solution of the following inequality?

$$\frac{x-1}{x+1} > 2x$$

Difficulty level   ○ Easy   ○ Normal   ● Hard

Calculation amount   ○ Small   ● Normal   ○ Large

1) $x < -1$

2) $x > -1$

3) $-1 < x < 1$

4) $-2 < x < -1$

Exercise: Solve the inequality equation below:

$$\frac{2x-1}{x+1} \le 1$$

1) $-1 \le x \le 2$

2) $-1 \le x < 2$

3) $-1 < x < 2$

4) $-1 < x \le 2$

Final answer: Choice (4).

5.24. How many distinctive real roots does the equation $(x^2 - 2x)^2 - (x^2 - 2x) = 2$ have?

Difficulty level   ○ Easy   ○ Normal   ● Hard

Calculation amount   ○ Small   ● Normal   ○ Large

1) 1

2) 2

3) 3

4) 4

5.25. For what value of m, does the quadratic equation $mx^2 + 5x + m^2 - 6 = 0$ have two real roots that are reciprocal to each other?

Difficulty level   ○ Easy   ○ Normal   ● Hard
Calculation amount   ○ Small   ○ Normal   ● Large
1) −2, 3
2) −2
3) 2
4) 3

> **Exercise:** For what value of m, does the quadratic equation $x^2 + 5x + m^2 = 0$ have two real roots that are reciprocal to each other?
> 1) 1
> 2) −1
> 3) ±1
> 4) 0
>
> *Final answer*: Choice (3).

5.26. For what value of m, the sum of the squares of the roots $2x^2 - mx + m - 1 = 0$ is 4?

Difficulty level   ○ Easy   ○ Normal   ● Hard
Calculation amount   ○ Small   ○ Normal   ● Large
1) −2, 6
2) −2
3) 2
4) 6

> **Exercise:** For what value of m, the sum of the squares of the roots $x^2 - mx + m - 1 = 0$ is 1?
> 1) 0
> 2) 1
> 3) 2
> 4) 3
>
> *Final answer*: Choice (2).

Reference

1. Rahmani-Andebili, M. (2021). Precalculus – Practice Problems, Methods, and Solutions, Springer Nature, 2021.

Solutions to Problems: Quadratic Equations

Abstract

In this chapter, the problems of the fifth chapter are fully solved, in detail, step-by-step, and with different methods.

6.1. Based on the information given in the problem, two times a positive quantity is 9 units fewer than one-third of its square value. In other words [1]:

$$x > 0 \tag{1}$$

$$2x = \frac{1}{3}x^2 - 9 \tag{2}$$

$$\Rightarrow x^2 - 6x - 27 = 0 \Rightarrow (x-9)(x+3) = 0 \Rightarrow x = -3, 9 \tag{3}$$

Solving (1) and (3):

$$x = 9$$

Choice (1) is the answer.

6.2. The value of the quadratic equation $ax^2 + bx + c$ is always positive, if $a > 0$ and $\Delta < 0$.

Therefore, for the given quadratic equation $ax^2 + 2x + 4a$, we can write:

$$a > 0 \tag{1}$$

$$\Delta = 2^2 - 4(a)(4a) < 0 \Rightarrow 4 - 16a^2 < 0 \Rightarrow a^2 > \frac{1}{4} \Rightarrow a > \frac{1}{2} \& \ a < -\frac{1}{2} \tag{2}$$

Solving (1) and (2):

$$a > \frac{1}{2}$$

Choice (1) is the answer.

M. Rahmani-Andebili, *Precalculus*, https://doi.org/10.1007/978-3-031-49364-5_6

6.3. Based on the information given in the problem, we have:

$$2x^2 + 3x - k = 0 \tag{1}$$

$$2X^2 - 5X + 1 = 0 \tag{2}$$

$$x = X - 2 \Rightarrow X = x + 2 \tag{3}$$

Solving (2) and (3):

$$2(x + 2)^2 - 5(x + 2) + 1 = 0 \Rightarrow 2x^2 + 3x - 1 = 0 \tag{4}$$

Comparing (1) and (4):

$$k = 1$$

Choice (1) is the answer.

6.4. The product of the roots of a quadratic equation can be determined as follows:

$$ax^2 + bx + c = 0 \Rightarrow \text{Product of roots} = \frac{c}{a}$$

Since the roots are reciprocal to each other, the product of the roots is equal to one. Hence:

$$2x^2 - 5x + m = 0 \Rightarrow \text{Product of roots} = \frac{m}{2} = 1 \Rightarrow m = 2$$

Choice (2) is the answer.

6.5. A quadratic equation has two real roots when its discriminant is equal to or greater than zero. In other words:

$$ax^2 + bx + c = 0 \Rightarrow \Delta = b^2 - 4ac \geq 0$$

Therefore, for the quadratic equation $(m + 2)x^2 + 4x + m - 1 = 0$, we can write:

$$\Delta = 4^2 - 4(m + 2)(m - 1) \geq 0 \Rightarrow 16 - 4(m^2 + m - 2) \geq 0 \Rightarrow m^2 + m - 6 \leq 0$$

$$\Rightarrow (m + 3)(m - 2) \leq 0 \Rightarrow -3 \leq m \leq 2$$

Choice (4) is the answer.

6.6. The value of the quadratic equation $ax^2 + bx + c$ is always negative, if $a < 0$ and $\Delta < 0$.

Therefore, for the given quadratic equation $m^3 x^2 + mx + \frac{1}{m}$, we can write:

$$m^3 < 0 \Rightarrow m < 0 \tag{1}$$

$$\Delta = m^2 - 4(m^3)\left(\frac{1}{m}\right) < 0 \Rightarrow m^2 - 4m^2 < 0 \Rightarrow -3m^2 < 0 \Rightarrow m^2 > 0 \Rightarrow m \in \mathbb{R} \tag{2}$$

Solving (1) and (2):

$$m < 0$$

Choice (2) is the answer.

6.7. Based on the information given in the problem, we know that:

$$X = 9x \Rightarrow x = \frac{X}{9} \tag{1}$$

$$x^2 + x - 3 = 0 \tag{2}$$

Solving (1) and (2):

$$\left(\frac{X}{9}\right)^2 + \frac{X}{9} - 3 = 0 \Rightarrow X^2 + 9X - 243 = 0$$

Choice (1) is the answer.

6.8. We know that a quadratic equation has the form below, where S and P are the sum and the product of its roots.

$$x^2 - Sx + P = 0, \quad S = \text{sum of roots} = -\frac{b}{a}, \quad P = \text{product of roots} = \frac{c}{a} \tag{1}$$

Based on the information given in the problem, we know that:

$$x_1 + x_2 = \frac{1}{x_1} \times \frac{1}{x_2} = \frac{1}{x_1 x_2} \tag{2}$$

Solving (1) and (2):

$$-\frac{b}{a} = \frac{1}{\frac{c}{a}} \Rightarrow -\frac{b}{a} = \frac{a}{c} \Rightarrow a^2 = -bc \Rightarrow a^2 + bc = 0$$

Choice (1) is the answer.

6.9. A quadratic equation has two real distinct roots, when its discriminant is greater than zero. In other words:

$$ax^2 + bx + c = 0 \Rightarrow \Delta = b^2 - 4ac > 0$$

Therefore:

$$2x^2 + ax + a - \frac{3}{2} = 0 \Rightarrow \Delta = a^2 - 4 \times 2 \times \left(a - \frac{3}{2}\right) > 0 \Rightarrow a^2 - 8a + 12 > 0$$

$$\Rightarrow (a - 6)(a - 2) > 0 \Rightarrow a < 2 \,\&\, a > 6$$

Choice (1) is the answer.

6.10. The value of the quadratic equation $ax^2 + bx + c$ is always nonnegative, if $a > 0$ and $\Delta \leq 0$.

Therefore, for the given quadratic equation $4x^2 - 2mx + 4m^2 \geq 0$, we can write:

$$4 > 0 \tag{1}$$

$$\Delta = (-2m)^2 - 4(4)(4m^2) \leq 0 \Rightarrow 4m^2 - 64m^2 \leq 0 \Rightarrow -60m^2 \leq 0 \Rightarrow m^2 \geq 0 \tag{2}$$

As can be seen in (1) and (2), both criteria are satisfied disregarding the value of m. Therefore:

$$m \in \mathbb{R}$$

Choice (1) is the answer.

6.11. Based on the information given in the problem, we have:

$$\frac{1}{x_1} + \frac{1}{x_2} = \frac{7}{3} \Rightarrow \frac{x_1 + x_2}{x_1 x_2} = \frac{7}{3} \tag{1}$$

As we know, the sum and the product of the roots of a quadratic equation can be determined as follows:

$$ax^2 + bx + c = 0 \Rightarrow S = \text{sum of roots} = -\frac{b}{a}, \quad P = \text{product of roots} = \frac{c}{a}$$

Hence, for the quadratic equation $2x^2 + (2k - 1)x - k = 0$, we can write:

$$x_1 + x_2 = -\frac{(2k-1)}{2} \tag{2}$$

$$x_1 x_2 = \frac{-k}{2} \tag{3}$$

Solving (1), (2), and (3):

$$\frac{-\frac{(2k-1)}{2}}{\frac{-k}{2}} = \frac{7}{3} \Rightarrow \frac{2k-1}{k} = \frac{7}{3} \Rightarrow 6k - 3 = 7k \Rightarrow k = -3$$

Choice (2) is the answer.

6.12. A quadratic equation has the following form, where S and P are the sum and the product of its roots.

$$x^2 - Sx + P = 0, \quad S = \text{sum of roots}, \quad P = \text{product of roots} \tag{1}$$

Therefore, for the given roots $2 + \sqrt{4 - a}$ and $2 - \sqrt{4 - a}$, we can write:

$$S = \left(2 + \sqrt{4 - a}\right) + \left(2 - \sqrt{4 - a}\right) = 4 \tag{2}$$

$$P = \left(2 + \sqrt{4 - a}\right)\left(2 - \sqrt{4 - a}\right) = 4 - (4 - a) = a \tag{3}$$

Solving (1), (2), and (3):

$$x^2 - 4x + a = 0$$

Choice (1) is the answer.

6.13. Based on the information given in the problem, we know that:

$$x_1^2 + x_2^2 = \frac{37}{4} \Rightarrow (x_1 + x_2)^2 - 2x_1x_2 = \frac{37}{4} \tag{1}$$

On the other hand, we know that a quadratic equation has the form below, where S and P are the sum and the product of its roots.

$$x^2 - Sx + P = 0, \quad S = \text{sum of roots} = -\frac{b}{a}, \quad P = \text{product of roots} = \frac{c}{a} \tag{2}$$

Solving (1) and (2):

$$S^2 - 2P = \frac{37}{4} \Rightarrow \left(-\frac{b}{a}\right)^2 - 2\left(\frac{c}{a}\right) = \frac{37}{4} \tag{3}$$

Solving (3) for the given quadratic equation $2x^2 - 5x + m = 0$:

$$\left(-\frac{-5}{2}\right)^2 - 2\frac{m}{2} = \frac{37}{4} \Rightarrow \frac{25}{4} - m = \frac{37}{4} \Rightarrow m = -3$$

Choice (2) is the answer.

6.14. The sum and the product of the roots of a quadratic equation can be determined as follows:

$$ax^2 + bx + c = 0 \Rightarrow S = \text{sum of roots} = -\frac{b}{a}, \quad P = \text{product of roots} = \frac{c}{a} \tag{1}$$

The problem can be solved as follows:

$$x_1^2 x_2 + x_1 x_2^2 = x_1 x_2 (x_1 + x_2) \tag{2}$$

Solving (1) and (2):

$$x_1^2 x_2 + x_1 x_2^2 = P \times S = \left(-\frac{b}{a}\right)\left(\frac{c}{a}\right) \tag{3}$$

For the quadratic equation $mnx^2 + n^2x + m^2 = 0$, we can write:

$$x_1^2 x_2 + x_1 x_2^2 = \left(-\frac{n^2}{mn}\right)\left(\frac{m^2}{mn}\right) = -1$$

Choice (1) is the answer.

6.15. Based on the information given in the problem, we have:

$$x_1 = -x_2 \Rightarrow x_1 + x_2 = 0 \tag{1}$$

As we know, the sum and the product of the roots of a quadratic equation are determined as follows:

$$ax^2 + bx + c = 0 \Rightarrow S = \text{sum of roots} = -\frac{b}{a}, \quad P = \text{product of roots} = \frac{c}{a} \tag{2}$$

Solving (1) and (2):

$$S = -\frac{b}{a} = 0 \Rightarrow b = 0 \tag{3}$$

Hence, for the given quadratic equation $(m + 1)x^2 + m(m^2 - 9)x - 2 = 0$, we can write:

$$m(m^2 - 9) = 0 \Rightarrow m = 0, -3, 3 \tag{4}$$

On the other hand, to have real roots for the given quadratic equation, the constraint below must be satisfied.

$$\Delta \geq 0 \Rightarrow m^2(m^2 - 9)^2 + 8(m + 1) \geq 0 \Rightarrow 8(m + 1) \geq 0 \Rightarrow m \geq -1 \tag{5}$$

Solving (4) and (5):

$$\Rightarrow m = 0, 3$$

However, $m = 0$ does not exist in the choices. Therefore, $m = 3$ is the only acceptable choice. Choice (3) is the answer.

6.16. The problem can be solved as follows:

$$x^2 - x + 2 = x^2 - x + \frac{1}{4} + \frac{7}{4} = x^2 - 2\left(\frac{1}{2}\right)x + \left(\frac{1}{2}\right)^2 + \frac{7}{4} = \left(x - \frac{1}{2}\right)^2 + \frac{7}{4}$$

The term $\left(x - \frac{1}{2}\right)^2$ is equal to or greater than zero. Hence:

$$\left(x - \frac{1}{2}\right)^2 \geq 0 \Rightarrow \left(x - \frac{1}{2}\right)^2 + \frac{7}{4} \geq \frac{7}{4} \Rightarrow \min\left\{\left(x - \frac{1}{2}\right)^2 + \frac{7}{4}\right\} = \frac{7}{4}$$

Choice (4) is the answer.

6.17. Based on the information given in the problem, we know that:

$$x_1^2 + x_2^2 = 12 \Rightarrow (x_1 + x_2)^2 - 2x_1x_2 = 12 \tag{1}$$

On the other hand, we know that a quadratic equation has the form below, where S and P are the sum and the product of its roots.

$$x^2 - Sx + P = 0, \quad S = \text{sum of roots} = -\frac{b}{a}, \quad P = \text{product of roots} = \frac{c}{a} \tag{2}$$

Solving (1) and (2):

$$S^2 - 2P = 12 \Rightarrow \left(-\frac{b}{a}\right)^2 - 2\left(\frac{c}{a}\right) = 12 \tag{3}$$

Solving (3) for the given quadratic equation $x^2 - 2kx - 2 = 0$:

$$\left(-\frac{-2k}{1}\right)^2 - 2\left(\frac{-2}{1}\right) = 12 \Rightarrow 4k^2 + 4 = 12 \Rightarrow k^2 = 2 \Rightarrow k = \pm\sqrt{2}$$

Choice (1) is the answer.

6.18. We know that a quadratic equation has the form below, where S and P are the sum and the product of its roots.

$$x^2 - Sx + P = 0, \quad S = \text{sum of roots} = -\frac{b}{a}, \quad P = \text{product of roots} = \frac{c}{a}$$

Therefore, for the given quadratic equation $2x^2 - 3x + m = 0$, we can write:

$$x_1 + x_2 = S = -\frac{-3}{2} = \frac{3}{2}$$

Based on the information given in the problem, we have:

$$x_1 - x_2 = \frac{5}{2}$$

Hence:

$$\Rightarrow \begin{cases} x_1 + x_2 = \dfrac{3}{2} \\ x_1 - x_2 = \dfrac{5}{2} \end{cases} \xrightarrow{+} 2x_1 = 4 \Rightarrow x_1 = 2$$

$$x_1 + x_2 = \frac{3}{2} \xrightarrow{x_1 = 2} x_2 = -\frac{1}{2}$$

$$x_1 x_2 = P = \frac{c}{a} \Rightarrow 2 \times \left(-\frac{1}{2}\right) = \frac{m}{2} \Rightarrow m = -2$$

Choice (1) is the answer.

6.19. Based on the information given in the problem, x_1 and x_2 are the roots of the quadratic equation $x^2 + ax - a^2 = 0$.

To determine the quadratic equation when its roots are $y_1 = x_1 + 1$ and $y_2 = x_2 + 1$, we need to replace x by $y - 1$ in the same quadratic equation, as follows:

$$(y - 1)^2 + a(y - 1) - a^2 = 0 \Rightarrow y^2 - 2y + 1 + ay - a - a^2 = 0$$

$$\Rightarrow y^2 + (a - 2)y + 1 - a - a^2 = 0 \xrightarrow{y \to x} x^2 + (a - 2)x + 1 - a - a^2 = 0$$

Choice (3) is the answer.

6.20. Based on the information given in the problem, we know that the sum of the squares of the two successive numbers (e.g., n and $n + 1$) is 925. In other words:

$$n^2 + (n + 1)^2 = 925 \Rightarrow n^2 + n^2 + 2n + 1 = 925 \Rightarrow 2n^2 + 2n = 924 \Rightarrow n(n + 1) = 462 \tag{1}$$

We can consider 462 as follows.

$$462 = 21 \times 22 \tag{2}$$

Solving (1) and (2):

$$n(n + 1) = 21 \times 22 \Rightarrow n = 21$$

$$\Rightarrow \text{The sum of the two numbers} = n + (n + 1) = 21 + 22 = 43$$

Choice (2) is the answer.

6.21. The sum and the product of the roots of a quadratic equation can be determined as follows:

$$ax^2 + bx + c = 0 \Rightarrow S = \text{sum of roots} = -\frac{b}{a}, \quad P = \text{product of roots} = \frac{c}{a} \tag{1}$$

The problem can be solved as follows:

$$\left(\sqrt{x'} - \sqrt{x''}\right)^2 = x' + x'' - 2\sqrt{x'x''} \tag{2}$$

$$x^2 - 4x + 1 = 0 \Rightarrow S = -\frac{b}{a} = -\frac{-4}{1} = 4, P = \frac{c}{a} = \frac{1}{1} = 1 \tag{3}$$

Solving (1), (2), and (3):

$$\left(\sqrt{x'} - \sqrt{x''}\right)^2 = x' + x'' - 2\sqrt{x'x''} = S - 2\sqrt{P} = 4 - 2\sqrt{1} = 2 \Rightarrow \left|\sqrt{x'} - \sqrt{x''}\right| = \sqrt{2}$$

Choice (1) is the answer.

6.22. Based on the information given in the problem, we have:

$$a > 0 \tag{1}$$

If it is assumed that x_1 is the common root, then we can write:

$$\begin{cases} x_1^2 + 2x_1 + a = 0 \\ x_1^2 - x_1 - 2a = 0 \end{cases} \xrightarrow{-} 3x_1 + 3a = 0 \Rightarrow x_1 = -a \tag{2}$$

By replacing $x_1 = -a$ in one of the equations, we have:

$$x_1^2 + 2x_1 + a = 0 \xrightarrow{x_1 = -a} (-a)^2 + 2(-a) + a = 0 \Rightarrow a^2 - a = 0 \Rightarrow a(a - 1) = 0 \Rightarrow a = 0, 1 \tag{3}$$

Solving (1) and (3):

$$a = 1 \overset{(2)}{\Rightarrow} x_1 = -1$$

Choice (2) is the answer.

6.23. The problem can be solved as follows:

$$\frac{x-1}{x+1} > 2x \Rightarrow \frac{x-1}{x+1} - 2x > 0 \Rightarrow \frac{x-1-2x^2-2x}{x+1} > 0 \Rightarrow \frac{2x^2+x+1}{x+1} < 0$$

The term $2x^2 + x + 1$ is always positive, because $a > 0$ and $\Delta < 0$, as can be seen below:

$$a = 2 > 0$$

$$\Delta = 1 - 4 \times 2 \times 1 = -7 < 0$$

Therefore:

$$x + 1 < 0 \Rightarrow x < -1$$

Choice (1) is the answer.

6.24. **First Method**: By defining the new variable below, the equation $(x^2 - 2x)^2 - (x^2 - 2x) = 2$ can be simplified and changed to a quadratic equation as follows:

$$X \equiv x^2 - 2x$$

$$\Rightarrow X^2 - X - 2 = 0$$

$$\Rightarrow X_1, X_2 = \frac{-b \pm \sqrt{b^2 - 4ac}}{2a} = \frac{-(-1) \pm \sqrt{(-1)^2 - 4 \times 1 \times (-2)}}{2 \times 1} = \frac{1 \pm \sqrt{9}}{2} = \frac{1 \pm 3}{2}$$

$$\Rightarrow X_1 = \frac{1+3}{2}, X_2 = \frac{1-3}{2} \Rightarrow X_1 = 2, X_2 = -1$$

Now, we need to come back to the primary variable (x). Therefore, we have two separate quadratic equations.

$$x^2 - 2x = 2$$

$$x^2 - 2x = -1$$

To solve $x^2 - 2x - 2 = 0$:

$$x_1, x_2 = \frac{-b \pm \sqrt{b^2 - 4ac}}{2a} = \frac{-(-2) \pm \sqrt{(-2)^2 - 4 \times 1 \times (-2)}}{2 \times 1} = \frac{2 \pm \sqrt{12}}{2} = 1 \pm \sqrt{3}$$

$$\Rightarrow x_1 = 1 + \sqrt{3}, x_2 = 1 - \sqrt{3}$$

As can be seen, this equation includes two distinctive real roots.

To solve $x^2 - 2x + 1 = 0$, we do not need to apply the quadratic equation formula, since it is a binomial expression.

$$(x-1)^2 = 0 \Rightarrow (x-1) = 0 \Rightarrow x = 1 \text{ (Double root)} \Rightarrow x_3 = 1, x_4 = 1$$

As can be seen, this equation has only one distinctive real root.

Therefore, the main equation has three distinctive real roots, overall. Choice (3) is the answer.

Second Method: In this method, we do not need to solve the quadratic equations. In fact, we can determine the number of distinctive real roots based on the value of discriminant of each quadratic equation ($\Delta = b^2 - 4ac$).

Regarding the equation $x^2 - 2x - 2 = 0$:

$$\Delta = b^2 - 4ac = (-2)^2 - 4 \times 1 \times (-2) = 12 > 0$$

Therefore, we will have two distinctive real roots, since the discriminant is greater than zero.

Moreover, regarding the equation $x^2 - 2x + 1 = 0$:

$$\Delta = b^2 - 4ac = (-2)^2 - 4 \times 1 \times 1 = 0$$

Thus, we will have just one distinctive real root, as the discriminant is equal to zero.

Overall, we will have three distinctive real roots. Choice (3) is the answer.

6.25. A quadratic equation has two real distinct roots when its discriminant is greater than zero. In other words:

$$ax^2 + bx + c = 0 \Rightarrow \Delta = b^2 - 4ac > 0$$

Therefore:

$$mx^2 + 5x + m^2 - 6 = 0 \Rightarrow \Delta = 5^2 - 4m(m^2 - 6) > 0 \tag{1}$$

Moreover, the product of the roots of a quadratic equation can be determined as follows:

$$ax^2 + bx + c = 0 \Rightarrow \text{Product of roots} = \frac{c}{a}$$

Since the roots are reciprocal to each other, the product of the roots is equal to 1. Hence:

$$mx^2 + 5x + m^2 - 6 = 0 \Rightarrow \text{Product of roots} = \frac{m^2 - 6}{m} = 1 \Rightarrow m^2 - m - 6 = 0$$

$$\Rightarrow (m+2)(m-3) = 0 \Rightarrow m = -2, 3 \tag{2}$$

Solving (1) and (2) for $m = 3$:

$$\Delta = 5^2 - 4m(m^2 - 6) \xrightarrow{m=3} 5^2 - 4 \times 3(3^2 - 6) = 25 - 12 \times 3 = -11 < 0$$

Therefore, $m = 3$ is not acceptable.

Solving (1) and (2) for $m = -2$:

$$\Delta = 5^2 - 4m(m^2 - 6) \xrightarrow{m = -2} 5^2 - 4 \times (-2)\left((-2)^2 - 6\right) = 25 - 16 = 9 > 0$$

Thus, $m = -2$ is not acceptable. Choice (2) is the answer.

6.26. Based on the information given in the problem, we know that:

$$x_1^2 + x_2^2 = 4 \Rightarrow (x_1 + x_2)^2 - 2x_1x_2 = 4 \tag{1}$$

On the other hand, we know that a quadratic equation has the form below, where S and P are the sum and the product of its roots.

$$x^2 - Sx + P = 0, \quad S = \text{sum of roots} = -\frac{b}{a}, \quad P = \text{product of roots} = \frac{c}{a} \tag{2}$$

Solving (1) and (2):

$$S^2 - 2P = 4 \Rightarrow \left(-\frac{b}{a}\right)^2 - 2\left(\frac{c}{a}\right) = 4 \tag{3}$$

Solving (3) for the given quadratic equation $2x^2 - mx + m - 1 = 0$:

$$\left(-\frac{-m}{2}\right)^2 - 2\frac{m-1}{2} = 4 \Rightarrow \frac{m^2}{4} - m + 1 = 4 \Rightarrow \frac{m^2}{4} - m - 3 = 0 \Rightarrow m^2 - 4m - 12 = 0$$

$$\Rightarrow (m - 6)(m + 2) = 0 \Rightarrow m = 6, -2$$

$m = 6$ is not acceptable because the equation will not have real roots, as can be seen below:

$$2x^2 - mx + m - 1 = 0 \xrightarrow{m = 6} 2x^2 - 6x + 5 = 0 \Rightarrow \Delta = (-6)^2 - 4(2)(5) = 36 - 40 = -4 < 0$$

$m = -2$ is acceptable because the equation will have two real roots, as presented below:

$$2x^2 - mx + m - 1 = 0 \xrightarrow{m = -2} 2x^2 + 2x - 3 = 0 \Rightarrow \Delta = (2)^2 - 4(2)(-3) = 4 + 24 = 28 > 0$$

Choice (2) is the answer.

Reference

1. Rahmani-Andebili, M. (2021). Precalculus – Practice Problems, Methods, and Solutions, Springer Nature, 2021.

Abstract

In this chapter, the basic and advanced problems of functions, algebra of functions, and inverse functions are presented. To help students study the chapter in the most efficient way, the problems are categorized into different levels based on their difficulty (easy, normal, and hard) and calculation amounts (small, normal, and large). Moreover, the problems are ordered from the easiest, with the smallest computations, to the most difficult, with the largest calculations.

7.1. If $f\left(\frac{1}{x}\right) = \sqrt{\frac{2x-1}{x^2}}$ and $g(x) = 2\cos^2(x)$, calculate the value of $fog\left(\frac{\pi}{3}\right)$ [1].

Difficulty level ● Easy ○ Normal ○ Hard
Calculation amount ● Small ○ Normal ○ Large

1) 0
2) $\frac{1}{2}$
3) $\frac{\sqrt{3}}{2}$
4) 2

Exercise: Calculate the value of $fog\left(\frac{\pi}{6}\right)$ if $f\left(\frac{1}{x}\right) = \sqrt{\frac{x-1}{x^2}}$ and $g(x) = \sin^2(x)$.

1) $\frac{\sqrt{3}}{2}$
2) $\frac{\sqrt{3}}{4}$
3) $\frac{1}{2}$
4) 1

Final answer: Choice (2).

7.2. Determine the value of $fog(3)$, if $f(x) = \sqrt{x} - 2$ and $g(x) = x + 1$.

Difficulty level ● Easy ○ Normal ○ Hard
Calculation amount ● Small ○ Normal ○ Large

1) 0
2) 1
3) 2
4) 3

7.3. If $f(x) = 2x - 2$ and $g(x) = x^2 - 1$, solve the equation $fog(x) = 0$.

Difficulty level   ● Easy   ○ Normal   ○ Hard

Calculation amount ● Small ○ Normal ○ Large

1) $\pm\sqrt{2}$

2) ± 2

3) $\pm\sqrt{3}$

4) ± 3

Exercise: Solve the equation $fog(x) = 0$, if $f(x) = x + 1$ and $g(x) = x^3 - 1$.

1) 0

2) 1

3) 2

4) 3

Final answer: Choice (1).

7.4. In the function below, calculate the value of $f(-2) + f(2)$.

$$f(x) = \begin{cases} 2x^2 + 4 & x \geq 2 \\ [x] - 4 & x < 2 \end{cases}$$

Difficulty level   ● Easy   ○ Normal   ○ Hard

Calculation amount ● Small ○ Normal ○ Large

1) 8

2) 6

3) 10

4) 5

Exercise: Calculate the value of $f(-1.5) + f(2.5)$ if:

$$f(x) = \begin{cases} 2[x] + 1 & x \geq 0 \\ |x| - 1 & x < 0 \end{cases}$$

1) 2.5

2) 5

3) 6.5

4) 5.5

Final answer: Choice (4).

7.5. Determine the domain of $y = \sqrt{2 - x^2}$.

Difficulty level ● Easy ○ Normal ○ Hard
Calculation amount ● Small ○ Normal ○ Large

1) $x \leq -\sqrt{2}, x \geq \sqrt{2}$
2) $-\sqrt{2} \leq x \leq \sqrt{2}$
3) $x = 0$
4) $-\sqrt{2} < x < \sqrt{2}$

Exercise: Determine the domain of the function below.

$$f(x) = \frac{1}{\sqrt{1 - x^2}}$$

1) $x \leq -1, x \geq 1$
2) $-1 \leq x \leq 1$
3) $x < -1, x > 1$
4) $-1 < x < 1$

Final answer: Choice (4).

7.6. Calculate $fog(x)$, if $f(x) = 1 - x^2$ and $g(x) = \sin(x)$.

Difficulty level ● Easy ○ Normal ○ Hard
Calculation amount ● Small ○ Normal ○ Large

1) $\cos^2(x)$
2) $\cos(x)$
3) $\sin(1 - x^2)$
4) $\sin(\cos(x))$

7.7. What is the inverse function $f(x) = \frac{1}{x}$?

Difficulty level ● Easy ○ Normal ○ Hard
Calculation amount ● Small ○ Normal ○ Large

1) x
2) $\frac{1}{x}$
3) $-\frac{1}{x}$
4) $\sqrt{x}$

Exercise: Calculate the inverse function of the function below.

$$f(x) = \frac{1}{x + 1}$$

1) $x + 1$
2) $\frac{1}{x+1}$
3) $\frac{1-x}{x}$
4) $\frac{1+x}{x}$

Final answer: Choice (3).

7.8. Calculate $fof(x)$, if $f(x) = \dfrac{1-x}{1+x}$.

     Difficulty level   ● Easy   ○ Normal   ○ Hard

     Calculation amount   ○ Small   ● Normal   ○ Large

1) $\left(\dfrac{1+x}{1-x}\right)^2$

2) 1

3) x

4) $\left(\dfrac{1-x}{1+x}\right)^2$

Exercise: Calculate $fof(x)$, if $f(x) = x^2$.

1) 1
2) x^2
3) x^4
4) $2x^2$

Final answer: Choice (3).

7.9. Determine the inverse function of $f(x) = x^2 - 2x$

     Difficulty level   ○ Easy   ● Normal   ○ Hard

     Calculation amount   ● Small   ○ Normal   ○ Large

1) $1 + \sqrt{x+1}$
2) $1 - \sqrt{x+1}$
3) $1 + \sqrt{x-1}$
4) $1 - \sqrt{x-1}$

Exercise: Calculate the inverse function of $f(x) = x^2 + 2x$.

1) $-1 + \sqrt{x+1}$
2) $-1 - \sqrt{x+1}$
3) $1 + \sqrt{x-1}$
4) $1 - \sqrt{x-1}$

Final answer: Choice (1).

7.10. Determine the value of $f(x)$ if $f(x + 1) = x^2 - 2x + 1$.

     Difficulty level   ○ Easy   ● Normal   ○ Hard

     Calculation amount   ● Small   ○ Normal   ○ Large

1) $(x - 2)^2$
2) $(x - 1)^2$
3) $x^2 - 2x$
4) $(x + 2)^2$

Exercise: Calculate the value of $f(x)$, if $f(x-1) = x^2 - 2x + 1$.
1) $(x+2)^2$
2) $(x-1)^2$
3) x^2
4) $(x-2)^2$

Final answer: Choice (3).

7.11. Which one of the terms below is not a function?

Difficulty level   ○ Easy   ● Normal   ○ Hard
Calculation amount   ● Small   ○ Normal   ○ Large

1) $y^2 = x$
2) $y^3 = x$
3) $y = \sqrt{x^2}$
4) $y = \begin{cases} x^2 & x \geq 0 \\ 1 & x < 0 \end{cases}$

Exercise: Which one of the following mathematical expressions is a function?
1) $|y| = x$
2) $y^4 = x$
3) $y = |x|$
4) $y^2 = x^2$

Final answer: Choice (3).

7.12. If $f(\sqrt{x}) = x + \sqrt{x}$, calculate the value of $f(2) + f(1)$.

Difficulty level   ○ Easy   ● Normal   ○ Hard
Calculation amount   ● Small   ○ Normal   ○ Large

1) 6
2) 7
3) 8
4) 9

Exercise: Calculate the value of $f(2)$ if $f(x^2) = x^2 + x^4$.
1) 1
2) 2
3) 4
4) 6

Final answer: Choice (4).

7.13. Calculate the value of $f(f(0))$, if:

$$f(x) = \begin{cases} x^2 + 1 & x \geq 1 \\ 2x + 3 & x < 1 \end{cases}$$

Difficulty level   ○ Easy   ● Normal   ○ Hard
Calculation amount   ● Small   ○ Normal   ○ Large
1) 3
2) 5
3) 10
4) 26

Exercise: Calculate the value of $f(f(0))$, if:

$$f(x) = \begin{cases} x^3 + 3 & x \geq 0 \\ x + 3 & x < 0 \end{cases}$$

1) 6
2) 30
3) 0
4) 3

Final answer: Choice (2).

7.14. Determine the domain of the function below.

$$f(x) = \sqrt{\frac{1 - |x|}{1 + |x|}}$$

Difficulty level   ○ Easy   ● Normal   ○ Hard
Calculation amount   ● Small   ○ Normal   ○ Large
1) $\mathbb{R}$
2) $x \leq 1$
3) $x \geq 1$
4) $-1 \leq x \leq 1$

Exercise: Calculate the domain of the following function:

$$f(x) = \sqrt{\frac{1 + |x|}{1 - |x|}}$$

Difficulty level   ○ Easy   ● Normal   ○ Hard

Calculation amount   ● Small   ○ Normal   ○ Large

1) $-1 \leq x \leq 1$
2) $x \leq 1$
3) $x \geq 1$
4) $-1 < x < 1$

Final answer: Choice (4).

7.15. Determine the domain of the function below.

$$f(x) = \sqrt{\frac{x - 1}{x - 3}} + \sqrt{\frac{2 - x}{x}}$$

Difficulty level   ○ Easy   ● Normal   ○ Hard

Calculation amount   ● Small   ○ Normal   ○ Large

1) $(0, 1]$
2) $[0, 1]$
3) $(0, 2]$
4) $(1, 3)$

Exercise: Calculate the domain of the following function:

$$f(x) = \sqrt{\frac{-x + 1}{x - 3}}$$

1) $1 < x < 3$
2) $1 \leq x \leq 3$
3) $x \leq 1, x > 3$
4) $1 \leq x < 3$

Final answer: Choice (4).

7.16. Determine the domain of the function below.

$$f(x) = \frac{\sqrt{x}}{|x| - 1}$$

Difficulty level   ○ Easy   ● Normal   ○ Hard
Calculation amount   ● Small   ○ Normal   ○ Large

1) $\mathbb{R}$
2) $[0, \infty) - \{1\}$
3) $\mathbb{R} - \{1\}$
4) $[0, \infty)$

Exercise: Determine the domain of the function below.

$$f(x) = \frac{2x}{|x| - 2}$$

1) $[0, \infty) - \{2\}$
2) $[0, \infty) - \{-2, 2\}$
3) $\mathbb{R} - \{-2, 2\}$
4) $\mathbb{R} - \{2\}$

Final answer: Choice (3).

7.17. Determine the domain of the function below.

$$f(x) = \frac{\sqrt{x} + 1}{x\sqrt{x} + 1}$$

Difficulty level   ○ Easy   ● Normal   ○ Hard
Calculation amount   ● Small   ○ Normal   ○ Large
1) $\mathbb{R} - \{0\}$
2) $[1, \infty)$
3) $[0, \infty)$
4) $\mathbb{R} - \{-1, 0\}$

Exercise: Determine the domain of the following function:

$$f(x) = \frac{x^3 + 1}{x^2 + 1}$$

1) $\mathbb{R} - \{-1, 1\}$
2) $\mathbb{R} - \{1\}$
3) $\mathbb{R} - \{-1\}$
4) $\mathbb{R}$

Final answer: Choice (4).

7.18. Which number does not exist in the domain of the function below?

$$f(x) = \frac{1 - x}{4x + x^3}$$

Difficulty level   ○ Easy   ● Normal   ○ Hard
Calculation amount   ● Small   ○ Normal   ○ Large
1) -2
2) 1
3) 2
4) 0

Exercise: Which number(s) does/do not exist in the domain of the following function?

$$f(x) = \frac{x + 3}{x^2 + 4x + 3}$$

1) -1
2) -3
3) $-1, -3$
4) 1, 3

Final answer: Choice (3).

7.19. Determine $g(x)$, if $f(x) = 2x$ and $f(g(x)) = 2x + 2$.
Difficulty level   ○ Easy   ● Normal   ○ Hard
Calculation amount   ● Small   ○ Normal   ○ Large
1) $x - 1$
2) $x + 2$
3) $x + 1$
4) $x - 2$

Exercise: Determine $g(x)$, if $f(x) = x - 1$ and $f(g(x)) = 3x + 4$.
1) $3x + 5$
2) $3x + 3$
3) $3x - 3$
4) $3x - 5$

Final answer: Choice (1).

7.20. Determine $g(x)$, if $f(x) = x - 1$ and $f(g(x)) = x$.
Difficulty level   ○ Easy   ● Normal   ○ Hard
Calculation amount   ● Small   ○ Normal   ○ Large
1) $x + 1$
2) $x^2 - x$
3) $2x - 1$
4) $x^2 - 1$

7.21. Calculate the inverse function of $f(x) = \sqrt{1-x}$.

1) $f^{-1}(x) = 1 - x^2,\ x \geq 0$
2) $f^{-1}(x) = \frac{1}{\sqrt{1-x}}$
3) $f^{-1}(x) = \sqrt{1+x}$
4) $f^{-1}(x) = 1 - x^2$

Exercise: Calculate the inverse function of $f(x) = \sqrt[3]{1+x}$.

1) $f^{-1}(x) = -1 + x^3,\ x \geq 0$
2) $f^{-1}(x) = -1 + x^3$
3) $f^{-1}(x) = \frac{1}{\sqrt[3]{1+x}}$
4) $f^{-1}(x) = 1 - x^3$

Final answer: Choice (2).

7.22. What is the inverse function of $f(x) = \sin(x) - 2$.

1) $2\mathrm{arc}(\sin(x))$
2) $-2\mathrm{arc}(\sin(x))$
3) $\mathrm{arc}(\sin(x - 2))$
4) $\mathrm{arc}(\sin(x + 2))$

Exercise: What is the inverse function of $f(x) = \cos(x + 1) + 1$.

1) $-1 + \mathrm{arc}(\sin(x - 1))$
2) $1 + \mathrm{arc}(\cos(x - 1))$
3) $-1 + \mathrm{arc}(\cos(x + 1))$
4) $-1 + \mathrm{arc}(\cos(x - 1))$

Final answer: Choice (4).

7.23. Calculate the inverse function of $f \circ g(x)$, if $f(x) = 3x - 2$ and $g(x) = 2 + x$.

1) $\frac{1}{3}x - \frac{4}{3}$
2) $3x - 4$
3) $\frac{1}{3}x + \frac{4}{3}$
4) $3x + 4$

Exercise: Calculate the inverse function of $f \circ g(x)$, if $f(x) = g(x) = x - 1$.

1) $x + 2$
2) $x - 2$
3) $-x + 2$
4) $-x - 2$

Final answer: Choice (1).

7.24. Calculate the inverse function of $f(x) = x^3 + 3x^2 + 3x + 2$.

Difficulty level   ○ Easy   ● Normal   ○ Hard
Calculation amount   ● Small   ○ Normal   ○ Large

1) $1 - \sqrt[3]{x - 1}$
2) $1 - \sqrt[3]{x + 1}$
3) $-1 + \sqrt[3]{x - 1}$
4) $-1 - \sqrt[3]{x + 1}$

Exercise: Calculate the inverse function of $f(x) = x^3 + 3x^2 + 3x$.

1) $1 + \sqrt[3]{x + 1}$
2) $1 - \sqrt[3]{x + 1}$
3) $-1 - \sqrt[3]{x - 1}$
4) $-1 + \sqrt[3]{x + 1}$

Final answer: Choice (4).

7.25. If $f(x) + xf(-x) = x^2 + 1$, then what is the value of $f(2)$?

Difficulty level   ○ Easy   ● Normal   ○ Hard
Calculation amount   ○ Small   ● Normal   ○ Large

1) -1
2) -2
3) 3
4) 4

Exercise: Determine the value of $f(1)$, if $f(x) + xf(-x) = x$.

1) 1
2) 2
3) 3
4) 4

Final answer: Choice (1).

7.26. Calculate the value of $f(f(f(2 \cos (x))))$, if $f(x) = x^2 - 2$.

1) $2\sin^8(x)$
2) $2\cos^8(x)$
3) $2 \sin (8x)$
4) $2 \cos (8x)$

7.27. Determine the domain of the following function:

$$f\left(\frac{x-1}{x}\right) = \sqrt{2x - 1}$$

1) $[-1, 0)$
2) $[-1, 1]$
3) $[-1, 1)$
4) $[1, \infty)$

Exercise: Determine the domain of the following function:

$$f\left(\frac{x}{x+1}\right) = \sqrt{x+1}$$

1) $x \leq 1$
2) $-1 \leq x \leq 1$
3) $x < 1$
4) $-1 < x < 1$

Final answer: Choice (3).

7.28. Determine the domain of $f(x) = \sqrt{1 - \sqrt{x - 1}}$.

1) $x \geq 1$
2) $1 \leq x \leq 2$
3) $x \leq 2$
4) $\frac{5}{4} \leq x \leq \frac{7}{4}$

7.29. Determine the domain of the function $f(x) = \sqrt{|x| - 1} + \sqrt{|x| + 1}$

1) $\mathbb{R} - [-1, 1]$
2) $\mathbb{R}$
3) $[-1, 1]$
4) $\mathbb{R} - (-1, 1)$

7.30. Calculate the value of $f(3)$, if:

$$f\left(x + \frac{1}{x}\right) = x^2 + \frac{1}{x^2}$$

Difficulty level   ○ Easy   ● Normal   ○ Hard
Calculation amount   ○ Small   ● Normal   ○ Large

1) $\dfrac{28}{3}$

2) $\dfrac{1}{7}$

3) 7

4) $\dfrac{3}{28}$

7.31. For what value of a, the function $f(x) = |x + 2| + a|x - 2|$ is even?
Difficulty level   ○ Easy   ● Normal   ○ Hard
Calculation amount   ○ Small   ● Normal   ○ Large

1) -1

2) 0

3) 1

4) 2

Exercise: For what value of a, the function $f(x) = |x + 2| + a|x - 2|$ is odd?
1) -1
2) 0
3) 1
4) 2

Final answer: Choice (1).

7.32. The functions $f(x) = x^2 + (A - 1)x$ and $g(x) = (B + 2)x^2 + \sin(x)$ are even and odd, respectively. Calculate the value of $A + B$.
Difficulty level   ○ Easy   ● Normal   ○ Hard
Calculation amount   ○ Small   ○ Normal   ● Large

1) -2

2) -1

3) 1

4) 2

7.33. Which one of the following functions is odd?
Difficulty level   ○ Easy   ● Normal   ○ Hard
Calculation amount   ○ Small   ○ Normal   ● Large

1) $\text{arc}(\cos(x))$

2) $\sqrt{1 - x} - \sqrt{1 + x}$

3) $x^4 + x$

4) $x \sin(x)$

Exercise: Which one of the following functions is odd?
1) x^2
2) $\cos(x)$
3) $x^2 + x$
4) $\sin(x)$

Final answer: Choice (4).

7.34. Which one of the functions below is odd?

Difficulty level   ○ Easy   ● Normal   ○ Hard
Calculation amount   ○ Small   ○ Normal   ● Large
1) $|x - 1| + |x + 1|$
2) $\sin(|x|)$
3) $x^3 + x^2$
4) $|x - 1| - |x + 1|$

7.35. Which one of the functions below is even?

Difficulty level   ○ Easy   ● Normal   ○ Hard
Calculation amount   ○ Small   ○ Normal   ● Large
1) $|x - 1| + |x + 1| + |x|$
2) $(x + 1)^4$
3) $f^2(x) + \sqrt[3]{x - 1} = 0$
4) $f(x) = [x] + 1$

7.36. We know that $f(g(x)) = x^2 + \dfrac{1}{x^2} - 4$ and $g(x) = x - \dfrac{1}{x}$. Determine $f(x)$.

Difficulty level   ○ Easy   ○ Normal   ● Hard
Calculation amount   ● Small   ○ Normal   ○ Large
1) $x^2 - 4$
2) $x^2 - 2$
3) x^2
4) $x^2 + 2$

7.37. Calculate the value of $f(-f(x))$, if $f(x) = \begin{cases} x^2 + 1 & x > 0 \\ 1 & x \leq 0 \end{cases}$.

Difficulty level   ○ Easy   ○ Normal   ● Hard
Calculation amount   ● Small   ○ Normal   ○ Large
1) 1
2) $x + 1$
3) $x^2 + 1$
4) $(x^2 + 1)^2 + 1$

Exercise: Calculate the value of $f(f(x))$, if $f(x) = \begin{cases} x^2 + 1 & x > 0 \\ 1 & x \leq 0 \end{cases}$.

1) 1
2) $x + 1$
3) $x^2 + 1$
4) $(x^2 + 1)^2 + 1$

Final answer: Choice (3).

7.38. Determine the domain of the function below.

$$f(x) = \sqrt{\log\left(\frac{5x - x^2}{4}\right)}$$

Difficulty level   ○ Easy   ○ Normal   ● Hard
Calculation amount   ○ Small   ● Normal   ○ Large

1) $1 < x < 4$
2) $0 < x < 5$
3) $1 \leq x \leq 4$
4) $0 \leq x \leq 5$

Exercise: Determine the domain of the following function:

$$f(x) = \ln \frac{2x - x^2}{2}$$

Difficulty level   ○ Easy   ○ Normal   ● Hard
Calculation amount   ○ Small   ● Normal   ○ Large

1) $-1 < x < 1$
2) $-1 \leq x \leq 1$
3) $0 < x < 2$
4) $0 \leq x \leq 2$

Final answer: Choice (3).

7.39. Determine the domain of $f(x) = \sqrt{\log_x(x^2 + 9)}$.

Difficulty level   ○ Easy   ○ Normal   ● Hard
Calculation amount   ○ Small   ● Normal   ○ Large

1) $(-\infty, \infty)$
2) $(0, \infty)$
3) $[-3, 3]$
4) $x > 0 - \{1\}$

Exercise: Calculate the domain of $f(x) = \log_x(-x^2 + 1)$.

1) $-1 < x < 1$
2) $0 \le x \le 1$
3) $0 < x < 1$
4) $-1 \le x \le 1$

Final answer: Choice (3).

7.40. Calculate the range of the function $f(x) = 2x - 2[x] + 1$.

Difficulty level   ○ Easy   ○ Normal   ● Hard
Calculation amount   ○ Small   ● Normal   ○ Large

1) $[0, 2]$
2) $[1, 3)$
3) $[0, 2)$
4) $[0, 3]$

7.41. Calculate the range of $fog(x)$, if $f(x) = x^2 + 1$ and $g(x) = \sqrt{x - 1}$.

Difficulty level   ○ Easy   ○ Normal   ● Hard
Calculation amount   ○ Small   ● Normal   ○ Large

1) $[0, \infty)$
2) $[1, \infty)$
3) $[-1, \infty)$
4) $\mathbb{R}$

7.42. Calculate the range of $f(x) = \sqrt{x^2 - 2x + 3}$.

Difficulty level   ○ Easy   ○ Normal   ● Hard
Calculation amount   ○ Small   ● Normal   ○ Large

1) $\left[\sqrt{2}, \infty\right)$
2) $\left[\sqrt{3}, \infty\right)$
3) $[0, \infty)$
4) $[1, \infty)$

Exercise: Calculate the range of $f(x) = \sqrt{x^2 - 8x + 17}$.

1) $\left[\sqrt{17}, \infty\right)$
2) $[3, \infty)$
3) $[0, \infty)$
4) $[1, \infty)$

Final answer: Choice (4).

7.43. Which one of the following functions is equivalent to $f(x) = |x - 2|$?

Difficulty level ○ Easy ○ Normal ● Hard
Calculation amount ○ Small ○ Normal ● Large

1) $g_1(x) = \left| \frac{x^2 - 3x + 2}{x - 1} \right|$

2) $g_2(x) = \left| \frac{x^2 - 4}{x + 2} \right|$

3) $g_3(x) = \frac{(x - 2)^2}{|x - 2|}$

4) $g_4(x) = \frac{|6x - 12|}{6}$

Exercise: Which one of the following functions is equivalent to $f(x) = |x + 1|$?

1) $(x + 1)$

2) $\left| \frac{x^2 - 1}{x - 1} \right|$

3) $\frac{|2x + 2|}{2}$

4) $\frac{(x + 1)^2}{|x + 1|}$

Final answer: Choice (3).

Reference

1. Rahmani-Andebili, M. (2021). Precalculus – Practice Problems, Methods, and Solutions, Springer Nature, 2021.

Solutions to Problems: Functions, Algebra of Functions, and Inverse Functions

Abstract

In this chapter, the problems of the seventh chapter are fully solved, in detail, step-by-step, and with different methods.

8.1. Based on the information given in the problem, we have [1]:

$$f\left(\frac{1}{x}\right) = \sqrt{\frac{2x-1}{x^2}}$$

$$g(x) = 2\cos^2(x)$$

The problem can be solved as follows:

$$\Rightarrow fog\left(\frac{\pi}{3}\right) = f\left(g\left(\frac{\pi}{3}\right)\right) = f\left(2\cos^2\left(\frac{\pi}{3}\right)\right) = f\left(2\left(\frac{1}{2}\right)^2\right) = f\left(\frac{1}{2}\right) = \sqrt{\frac{2(2)-1}{2^2}} = \frac{\sqrt{3}}{2}$$

Choice (3) is the answer.

8.2. Based on the information given in the problem, we have:

$$f(x) = \sqrt{x} - 2$$

$$g(x) = x + 1$$

The problem can be solved as follows:

$$\Rightarrow fog(3) = f(g(3)) = f(3+1) = f(4) = \sqrt{4} - 2 = 0$$

Choice (1) is the answer.

8.3. Based on the information given in the problem, we have:

$$f(x) = 2x - 2$$

$$g(x) = x^2 - 1$$

The problem can be solved as follows:

$$\Rightarrow fog(x) = f(g(x)) = f(x^2 - 1) = 2(x^2 - 1) - 2 = 2x^2 - 4$$

$$fog(x) = 0 \Rightarrow 2x^2 - 4 = 0 \Rightarrow x^2 = 2 \Rightarrow x = \pm\sqrt{2}$$

Choice (1) is the answer.

8.4. Based on the information given in the problem, we have:

$$f(x) = \begin{cases} 2x^2 + 4 & x \geq 2 \\ [x] - 4 & x < 2 \end{cases}$$

The problem can be solved as follows:

$$\Rightarrow f(2) = 2(2)^2 + 4 = 12$$

$$\Rightarrow f(-2) = [-2] - 4 = -6$$

$$\Rightarrow f(2) + f(-2) = 12 + (-6) = 6$$

Choice (2) is the answer.

8.5. Based on the information given in the problem, we have:

$$y = \sqrt{2 - x^2}$$

The domain of a function in radical form, including even root, is determined by considering the radicand equal to and greater than zero. Therefore:

$$2 - x^2 \geq 0 \Rightarrow x^2 \leq 2 \Rightarrow -\sqrt{2} \leq x \leq \sqrt{2}$$

Choice (2) is the answer.

8.6. From trigonometry, we know that:

$$\sin^2(x) + \cos^2(x) = 1$$

Based on the information given in the problem, we have:

$$f(x) = 1 - x^2$$

$$g(x) = \sin(x)$$

Therefore:

$$fog(x) = f(g(x)) = f(\sin(x)) = 1 - (\sin(x))^2 = \cos^2(x)$$

Choice (1) is the answer.

8.7. Based on the information given in the problem, we have:

$$f(x) = \frac{1}{x}$$

To determine the inverse of a function, it is convenient to use y instead of $f(x)$. Then, we need to determine x based on y. After that, we must replace x with y and vice versa. Note that the domain of $f^{-1}(x)$ is the same as the range of $f(x)$.

Therefore:

$$y = \frac{1}{x} \Rightarrow x = \frac{1}{y} \Rightarrow y = \frac{1}{x} \overset{or}{\Rightarrow} f^{-1}(x) = \frac{1}{x}$$

Choice (2) is the answer.

8.8. Based on the information given in the problem, we have:

$$f(x) = \frac{1-x}{1+x}$$

Therefore:

$$fof(x) = f(f(x)) = f\left(\frac{1-x}{1+x}\right) = \frac{1 - \frac{1-x}{1+x}}{1 + \frac{1-x}{1+x}} = \frac{\frac{1+x-(1-x)}{1+x}}{\frac{1+x+1-x}{1+x}} = \frac{\frac{2x}{1+x}}{\frac{2}{1+x}} = x$$

Choice (3) is the answer.

8.9. Based on the information given in the problem, we have:

$$f(x) = x^2 - 2x$$

First, we need to define the function in a square form as follows:

$$\Rightarrow f(x) = x^2 - 2x + 1 - 1 \Rightarrow f(x) = (x-1)^2 - 1$$

To determine the inverse of a function, it is convenient to use y instead of $f(x)$. Then, we need to determine x based on y. After that, we must replace x with y and vice versa. Note that the domain of $f^{-1}(x)$ is the same as the range of $f(x)$.

Therefore:

$$\Rightarrow y = (x-1)^2 - 1 \Rightarrow y + 1 = (x-1)^2 \Rightarrow \sqrt{y+1} = x - 1 \Rightarrow \sqrt{y+1} + 1 = x \Rightarrow y = \sqrt{y+1} + 1$$

$$\overset{or}{\Rightarrow} f^{-1}(x) = 1 + \sqrt{y+1}$$

Choice (1) is the answer.

8.10. Based on the information given in the problem, we have:

$$f(x+1) = x^2 - 2x + 1$$

The problem can be solved as follows:

$$\xRightarrow{x \to x-1} f((x-1)+1) = (x-1)^2 - 2(x-1) + 1 \Rightarrow f(x) = x^2 - 2x + 1 - 2x + 2 + 1$$

$$\Rightarrow f(x) = x^2 - 4x + 4 \Rightarrow f(x) = (x-2)^2$$

Choice (1) is the answer.

8.11. A mathematical relation is a function if for any value of x, one value of y is achieved at most. Or, a function is a binary relation between two sets that associates every element of the first set to exactly one element of the second set. Herein, $y^2 = x$ is not a function because, for $x = 1$, $y = -1, 1$ are achieved. Choice (1) is the answer.

8.12. Based on the information given in the problem, we have:

$$f\left(\sqrt{x}\right) = x + \sqrt{x}$$

The problem can be solved as follows:

$$f\left(\sqrt{x}\right) = x + \sqrt{x} = \left(\sqrt{x}\right)^2 + \sqrt{x} \xRightarrow{\sqrt{x} \to x} f(x) = x^2 + x$$

$$\Rightarrow f(2) + f(1) = 2^2 + 2 + 1^2 + 1 = 8$$

Choice (3) is the answer.

8.13. Based on the information given in the problem, we have:

$$f(x) = \begin{cases} x^2 + 1 & x \geq 1 \\ 2x + 3 & x < 1 \end{cases}$$

The problem can be solved as follows:

$$f(0) = 2 \times 0 + 3 = 3 \Rightarrow f(f(0)) = f(3) = 3^2 + 1 = 10$$

Choice (3) is the answer.

8.14. Based on the information given in the problem, we have:

$$f(x) = \sqrt{\frac{1 - |x|}{1 + |x|}}$$

The domain of a function in radical form, including even root, is determined by considering the radicand equal to and greater than zero. Therefore:

$$\frac{1 - |x|}{1 + |x|} \geq 0 \xRightarrow{1 + |x| > 0} 1 - |x| \geq 0 \Rightarrow |x| \leq 1 \Rightarrow 1 \leq x \leq 1$$

Choice (4) is the answer.

8.15. Based on the information given in the problem, we have:

$$f(x) = \sqrt{\frac{x-1}{x-3}} + \sqrt{\frac{2-x}{x}}$$

The domain of a function in radical form, including even root, is determined by considering the radicand equal to and greater than zero. Therefore:

$$\begin{cases} \dfrac{x-1}{x-3} \geq 0 \\ \dfrac{2-x}{x} \geq 0 \end{cases} \Rightarrow \begin{cases} \dfrac{x-1}{x-3} \geq 0 \\ \dfrac{x-2}{x} \leq 0 \end{cases} \Rightarrow \begin{cases} x \leq 1, x > 3 \\ 0 < x \leq 2 \end{cases} \xRightarrow{\cap} 0 < x \leq 1$$

Note that $x = 0$ must be excluded from the domain, since it makes the denominator zero. Choice (1) is the answer.

8.16. Based on the information given in the problem, we have:

$$f(x) = \frac{\sqrt{x}}{|x| - 1}$$

The domain of a function in radical form, including even root, is determined by considering the radicand equal to and greater than zero. Moreover, the values of those x that make the denominator zero must be excluded from the domain. Thus:

$$\begin{cases} x \geq 0 \\ |x| - 1 \neq 0 \end{cases} \Rightarrow \begin{cases} x \geq 0 \\ |x| \neq 1 \end{cases} \Rightarrow \begin{cases} x \geq 0 \\ x \neq \pm 1 \end{cases} \xRightarrow{\cap} D_f = [0, \infty) - \{1\}$$

Choice (2) is the answer.

8.17. Based on the information given in the problem, we have:

$$f(x) = \frac{\sqrt{x} + 1}{x\sqrt{x} + 1}$$

The domain of a function in radical form, including even root, is determined by considering the radicand equal to and greater than zero. Moreover, the values of those x that make the denominator zero must be excluded from the domain. Thus:

$$\begin{cases} x \geq 0 \\ x\sqrt{x} + 1 \neq 0 \end{cases}$$

Note that $x\sqrt{x} + 1 \neq 0$ is true for any x. Hence:

$$\Rightarrow x \geq 0$$

Choice (3) is the answer.

8.18. Based on the information given in the problem, we have:

$$f(x) = \frac{1-x}{4x+x^3}$$

The values of those x that make the denominator zero are not in the domain. Thus:

$$4x + x^3 \neq 0 \Rightarrow x(4 + x^2) \neq 0$$

Note that $4 + x^2 \neq 0$ for any x. Hence:

$$\Rightarrow x = 0$$

Choice (4) is the answer.

8.19. Based on the information given in the problem, we have:

$$f(x) = 2x \tag{1}$$

$$f(g(x)) = 2x + 2 \tag{2}$$

Therefore:

$$f(g(x)) = 2g(x) \tag{3}$$

Solving (2) and (3):

$$2g(x) = 2x + 2 \Rightarrow g(x) = x + 1$$

Choice (3) is the answer.

8.20. Based on the information given in the problem, we have:

$$f(x) = x - 1 \tag{1}$$

$$f(g(x)) = x \tag{2}$$

Thus:

$$f(g(x)) = g(x) - 1 \tag{3}$$

Solving (2) and (3):

$$g(x) - 1 = x \Rightarrow g(x) = x + 1$$

Choice (1) is the answer.

8.21. Based on the information given in the problem, we have:

$$f(x) = \sqrt{1-x}$$

To determine the inverse of a function, it is convenient to use y instead of $f(x)$. Then, we need to determine x based on y. After that, we must replace x with y and vice versa. Note that the domain of $f^{-1}(x)$ is the same as the range of $f(x)$.

Therefore:

$$y = \sqrt{1-x} \Rightarrow y^2 = 1 - x \Rightarrow x = 1 - y^2 \Rightarrow y = 1 - x^2 \overset{\text{or}}{\Rightarrow} f^{-1}(x) = 1 - x^2$$

Since the domain of $f^{-1}(x)$ is the same as the range of $f(x)$, which is $[0, \infty)$, we need to add $x \geq 0$ to $f^{-1}(x)$ as its domain. Thus:

$$f^{-1}(x) = 1 - x^2, x \geq 0$$

Choice (1) is the answer.

8.22. Based on the information given in the problem, we have:

$$f(x) = \sin(x) - 2$$

To determine the inverse of a function, it is convenient to use y instead of $f(x)$. Then, we need to determine x based on y. After that, we must replace x with y and vice versa. Note that the domain of $f^{-1}(x)$ is the same as the range of $f(x)$.

Therefore:

$$y = \sin(x) - 2 \Rightarrow y + 2 = \sin(x) \Rightarrow x = \text{arc}(\sin(y + 2)) \Rightarrow y = \text{arc}(\sin(x + 2)) \overset{\text{or}}{\Rightarrow} f^{-1}(x) = \text{arc}(\sin(x + 2))$$

Choice (4) is the answer.

8.23. Based on the information given in the problem, we have:

$$f(x) = 3x - 2$$

$$g(x) = 2 + x$$

First, we need to determine $fog(x)$ as follows:

$$fog(x) = f(g(x)) = f(2 + x) = 3(2 + x) - 2 = 3x + 4$$

To determine the inverse of a function, it is convenient to use y instead of $f(x)$. Then, we need to determine x based on y. After that, we must replace x with y and vice versa. Note that the domain of $f^{-1}(x)$ is the same as the range of $f(x)$.

Therefore:

$$y = 3x + 4 \Rightarrow 3x = y - 4 \Rightarrow x = \frac{1}{3}y - \frac{4}{3} \Rightarrow y = \frac{1}{3}x - \frac{4}{3}$$

Choice (1) is the answer.

8.24. Based on the information given in the problem, we have:

$$f(x) = x^3 + 3x^2 + 3x + 2$$

First, we need to define the function in cube form as follows:

$$\Rightarrow f(x) = (x^3 + 3x^2 + 3x + 1) + 1 = (x + 1)^3 + 1$$

To determine the inverse of a function, it is convenient to use y instead of $f(x)$. Then, we need to determine x based on y. After that, we must replace x with y and vice versa. Note that the domain of $f^{-1}(x)$ is the same as the range of $f(x)$.

Therefore:

$$\Rightarrow y = (x+1)^3 + 1 \Rightarrow y - 1 = (x+1)^3 \Rightarrow (y-1)^{\frac{1}{3}} = x + 1 \Rightarrow x = (y-1)^{\frac{1}{3}} - 1 \Rightarrow y = (x-1)^{\frac{1}{3}} - 1$$

$$\overset{or}{\Rightarrow} f^{-1}(x) = -1 + \sqrt[3]{x-1}$$

Choice (3) is the answer.

8.25. Based on the information given in the problem, we have:

$$f(x) + xf(-x) = x^2 + 1$$

The problem can be solved as follows:

$$\begin{array}{l} \xrightarrow{x=2} \\ \xrightarrow{x=-2} \end{array} \left\{ \begin{array}{l} f(2) + 2f(-2) = 2^2 + 1 \\ f(-2) + (-2)f(2) = (-2)^2 + 1 \end{array} \right. \Rightarrow \left\{ \begin{array}{l} f(2) + 2f(-2) = 5 \\ f(-2) - 2f(2) = 5 \end{array} \right. \overset{\times(-2)}{\Rightarrow} \left\{ \begin{array}{l} f(2) + 2f(-2) = 5 \\ -2f(-2) + 4f(2) = -10 \end{array} \right.$$

$$\overset{+}{\Rightarrow} 5f(2) = -5 \Rightarrow f(2) = -1$$

Choice (1) is the answer.

8.26. From trigonometry, we know that:

$$1 + \cos(2x) = 2\cos^2(x)$$

Based on the information given in the problem, we have:

$$f(x) = x^2 - 2$$

The problem can be solved as follows:

$$\Rightarrow f(2\cos(x)) = (2\cos(x))^2 - 2 = 4\cos^2(x) - 2 = 2(1 + \cos(2x)) - 2 = 2\cos(2x)$$

$$\Rightarrow f(f(2\cos(x))) = (2\cos(2x))^2 - 2 = 4\cos^2(2x) - 2 = 2(1 + \cos(4x)) - 2 = 2\cos(4x)$$

$$\Rightarrow f(f(f(2\cos(x)))) = (2\cos(4x))^2 - 2 = 4\cos^2(4x) - 2 = 2(1 + \cos(8x)) - 2 = 2\cos(8x)$$

Choice (4) is the answer.

8.27. Based on the information given in the problem, we have:

$$f\left(\frac{x-1}{x}\right) = \sqrt{2x-1}$$

First, we need to determine $f(x)$ as follows:

$$\frac{x-1}{x} \equiv t \Rightarrow x = \frac{1}{1-t} \Rightarrow f(t) = \sqrt{2 \times \frac{1}{1-t} - 1} = \sqrt{\frac{1+t}{1-t}} \xrightarrow{t \to x} f(x) = \sqrt{\frac{1+x}{1-x}}$$

The domain of a function in radical form, including even root, is determined by considering the radicand equal to and greater than zero. Therefore:

$$\frac{1+x}{1-x} \geq 0 \Rightarrow \frac{x+1}{x-1} \leq 0 \Rightarrow -1 \leq x < 1$$

Note that $x = 1$ must be excluded from the domain, since it makes the denominator zero. Choice (3) is the answer.

8.28. Based on the information given in the problem, we have:

$$f(x) = \sqrt{1 - \sqrt{x-1}}$$

The domain of a function in radical form, including even root, is determined by considering the radicand equal to and greater than zero. Therefore:

$$\begin{cases} 1 - \sqrt{x-1} \geq 0 \\ x - 1 \geq 0 \end{cases} \Rightarrow \begin{cases} \sqrt{x-1} \leq 1 \\ x \geq 1 \end{cases} \Rightarrow \begin{cases} x - 1 \leq 1 \\ x \geq 1 \end{cases} \Rightarrow \begin{cases} x \leq 2 \\ x \geq 1 \end{cases} \stackrel{\cap}{\Longrightarrow} 1 \leq x \leq 2$$

Choice (2) is the answer.

8.29. Based on the information given in the problem, we have:

$$f(x) = \sqrt{|x| - 1} + \sqrt{|x| + 1}$$

The domain of a function in radical form, including even root, is determined by considering the radicand equal to and greater than zero.

$$\begin{cases} |x| - 1 \geq 0 \\ |x| + 1 \geq 0 \end{cases} \Rightarrow \begin{cases} |x| \geq 1 \\ x \in \mathbb{R} \end{cases} \Rightarrow \begin{cases} x \leq -1, x \geq 1 \\ x \in \mathbb{R} \end{cases} \stackrel{\cap}{\Rightarrow} x \leq -1, x \geq 1 \Rightarrow D_f = \mathbb{R} - (-1, 1)$$

Note that $|x| + 1 \geq 0$ is true for any x. Choice (4) is the answer.

8.30. Based on the information given in the problem, we have:

$$f\left(x + \frac{1}{x}\right) = x^2 + \frac{1}{x^2}$$

The problem can be solved as follows:

$$\Rightarrow f\left(x + \frac{1}{x}\right) = x^2 + \frac{1}{x^2} + 2 - 2 = \left(x + \frac{1}{x}\right)^2 - 2$$

$$\xrightarrow{x + \frac{1}{x} \to x} f(x) = x^2 - 2 \Rightarrow f(3) = 3^2 - 2 = 7$$

Choice (3) is the answer.

8.31. Based on the information given in the problem, we have:

$$f(x) = |x + 2| + a|x - 2| \tag{1}$$

Based on the definition, the function $f(x)$ is even if its domain is symmetric and:

$$f(-x) = f(x) \tag{2}$$

Therefore:

$$\Rightarrow f(-x) = |-x + 2| + a|-x - 2| = |-(x - 2)| + a|-(x + 2)| = |x - 2| + a|x + 2| \tag{3}$$

Solving (1), (2), and (3):

$$|x - 2| + a|x + 2| = |x + 2| + a|x - 2| \Rightarrow a = 1$$

Choice (3) is the answer.

8.32. Based on the information given in the problem, we have:

$$f(x) = x^2 + (A - 1)x \text{ is an even function} \tag{1}$$

$$g(x) = (B + 2)x^2 + \sin(x) \text{ is an odd function} \tag{2}$$

Based on the definition, the function $f(x)$ is even if its domain is symmetric and:

$$f(-x) = f(x) \tag{3}$$

Additionally, the function $f(x)$ is odd if its domain is symmetric and:

$$f(-x) = -f(x) \tag{4}$$

Solving (1) and (3):

$$(-x)^2 + (A - 1)(-x) = x^2 + (A - 1)x \Rightarrow 2(A - 1)x = 0 \Rightarrow A = 1$$

Solving (2) and (4):

$$(B + 2)(-x)^2 + \sin(-x) = -\left((B + 2)x^2 + \sin(x)\right)$$

$$\Rightarrow (B + 2)x^2 - \sin(x) = -(B + 2)x^2 - \sin(x) \Rightarrow 2(B + 2)x^2 = 0 \Rightarrow B = -2$$

Therefore:

$$A + B = 1 + (-2) = -1$$

Choice (2) is the answer.

8.33. Based on the definition, the function $f(x)$ is odd if its domain is symmetric and:

$$f(-x) = -f(x)$$

Choice (1):

$$f(x) = \text{arc}(\cos(x)) \Rightarrow f(-x) = \text{arc}(\cos(-x)) = \pi - \text{arc}(\cos(x)) \neq \{-f(x), f(x)\} \Rightarrow \text{Not even nor odd}$$

Choice (2):

$$f(x) = \sqrt{1-x} - \sqrt{1+x} \Rightarrow f(-x) = \sqrt{1-(-x)} - \sqrt{1+(-x)} = \sqrt{1+x} - \sqrt{1-x}$$
$$= -\left(\sqrt{1-x} - \sqrt{1+x}\right) = -f(x) \Rightarrow \text{Odd}$$

Choice (3):

$$f(x) = x^4 + x \Rightarrow f(-x) = (-x)^4 + (-x) = x^4 - x \neq \{-f(x), f(x)\} \Rightarrow \text{Not even nor odd}$$

Choice (4):

$$f(x) = x\sin(x) \Rightarrow f(-x) = -x\sin(-x) = x\sin(x) = f(x) \Rightarrow \text{Even}$$

Choice (2) is the answer.

8.34. Based on the definition, the function $f(x)$ is odd if its domain is symmetric and:

$$f(-x) = -f(x)$$

Choice (1):

$$f(x) = |x-1| + |x+1| \Rightarrow f(-x) = |-x-1| + |-x+1| = |-(x+1)| + |-(x-1)|$$

$$= |x+1| + |x-1| = f(x) \Rightarrow \text{Even}$$

Choice (2):

$$f(x) = \sin(|x|) \Rightarrow f(-x) = \sin(|-x|) = \sin(|x|) = f(x) \Rightarrow \text{Even}$$

Choice (3):

$$f(x) = x^3 + x^2 \Rightarrow f(-x) = (-x)^3 + (-x)^2 = -x^3 + x^2 \neq \{-f(x), f(x)\} \Rightarrow \text{Not even nor odd}$$

Choice (4):

$$f(x) = |x-1| - |x+1| \Rightarrow f(-x) = |-x-1| - |-x+1| = |-(x+1)| - |-(x-1)|$$
$$= |x+1| - |x-1| = -f(x) \Rightarrow \text{Odd}$$

Choice (4) is the answer.

8.35. Based on the definition, the function $f(x)$ is even if its domain is symmetric and:

$$f(-x) = f(x)$$

Choice (1):

$$f(x) = |x - 1| + |x + 1| + |x| \Rightarrow f(-x) = |-x - 1| + |-x + 1| + |-x|$$
$$= |-(x + 1)| + |-(x - 1)| + |-x| = |x + 1| + |x - 1| + |x| = f(x) \Rightarrow \text{Even}$$

Choice (2):

$$f(x) = (x + 1)^4 \Rightarrow f(-x) = (-x + 1)^4 = (x - 1)^4 \neq \{-f(x), f(x)\} \Rightarrow \text{Not even nor odd}$$

Choice (3):

$$f^2(x) + \sqrt[3]{x - 1} = 0 \Rightarrow \text{Not a function}$$

Choice (4):

$$f(x) = \lfloor x \rfloor + 1 \Rightarrow f(-x) = \lfloor -x \rfloor + 1 \neq \{-f(x), f(x)\} \Rightarrow \text{Not even nor odd}$$

Choice (1) is the answer.

8.36. Based on the information given in the problem, we have:

$$g(x) = x - \frac{1}{x} \tag{1}$$

$$f(g(x)) = x^2 + \frac{1}{x^2} - 4 \tag{2}$$

Therefore:

$$\Rightarrow f(g(x)) = x^2 + \frac{1}{x^2} - 2 - 2 = \left(x - \frac{1}{x}\right)^2 - 2 \tag{3}$$

Solving (1) and (3):

$$f(g(x)) = (g(x))^2 - 2 \Rightarrow f(x) = x^2 - 2$$

Choice (2) is the answer.

8.37. Based on the information given in the problem, we have:

$$f(x) = \begin{cases} x^2 + 1 & x > 0 \\ 1 & x \leq 0 \end{cases}$$

As can be noticed from $f(x)$, the value of function is always positive. Therefore, the value of $-f(x)$ is always negative. Hence:

$$f(-f(x)) = 1$$

Choice (1) is the answer.

8.38. Based on the information given in the problem, we have:

$$f(x) = \sqrt{\log\left(\frac{5x - x^2}{4}\right)}$$

The domain of a function in radical form, including even root, is determined by considering the radicand equal to and greater than zero. Moreover, the domain of a logarithmic function can be determined by considering its argument greater than zero. Therefore:

$$\begin{cases} \log\left(\dfrac{5x - x^2}{4}\right) \geq 0 \\ \dfrac{5x - x^2}{4} > 0 \end{cases} \Rightarrow \begin{cases} \log\left(\dfrac{5x - x^2}{4}\right) \geq \log(1) \\ x^2 - 5x < 0 \end{cases} \Rightarrow \begin{cases} \dfrac{5x - x^2}{4} \geq 1 \\ x(x - 5) < 0 \end{cases} \Rightarrow \begin{cases} x^2 - 5x + 4 \leq 0 \\ x(x - 5) < 0 \end{cases}$$

$$\Rightarrow \begin{cases} (x - 4)(x - 1) \leq 0 \\ x(x - 5) < 0 \end{cases} \Rightarrow \begin{cases} 1 \leq x \leq 4 \\ 0 < x < 5 \end{cases} \xRightarrow{\cap} 1 \leq x \leq 4$$

Choice (3) is the answer.

8.39. Based on the information given in the problem, we have:

$$f(x) = \sqrt{\log_x(x^2 + 9)}$$

The domain of a function in radical form, including even root, is determined by considering the radicand equal to and greater than zero. In addition, the domain of a logarithmic function can be determined by considering its argument greater than zero. In addition, the base of the logarithm must be greater than zero but not equal to one. Therefore: .

$$\begin{cases} \log_x(x^2 + 9) \geq 0 \\ x^2 + 9 > 0 \\ x > 0, x \neq 1 \end{cases} \Rightarrow \begin{cases} \log_x(x^2 + 9) \geq \log_x(1) \\ x^2 + 9 > 0 \\ x > 0, x \neq 1 \end{cases} \Rightarrow \begin{cases} x^2 + 9 \geq 1 \\ x^2 + 9 > 0 \\ x > 0, x \neq 1 \end{cases}$$

Note that $x^2 + 9 > 0$ and $x^2 + 8 \geq 0$ are true for any x. Hence:

$$\xRightarrow{\cap} x > 0, x \neq 1$$

Choice (4) is the answer.

8.40. Based on the information given in the problem, we have:

$$f(x) = 2x - 2[x] + 1$$

Based on the definition, we know that:

$$[x] \leq x < [x] + 1 \xRightarrow{-[x]} 0 \leq x - [x] < 1 \xRightarrow{\times 2} 0 \leq 2x - 2[x] < 2 \xRightarrow{+1} 1 \leq 2x - 2[x] + 1 < 3$$

$$\Rightarrow 1 \leq f(x) < 3 \Rightarrow R_f = [1, 3)$$

Choice (2) is the answer.

8.41. Based on the information given in the problem, we have:

$$f(x) = x^2 + 1$$

$$g(x) = \sqrt{x - 1}$$

$$\Rightarrow fog(x) = f(g(x)) = f\left(\sqrt{x-1}\right) = \left(\sqrt{x-1}\right)^2 + 1 = x - 1 + 1 = x$$

Next, we need to determine the domain of the function. As we know, the domain of a function in radical form, including even root, is determined by considering the radicand equal to and greater than zero.

$$x - 1 \geq 0 \Rightarrow x \geq 1 \Rightarrow D_{fog} = [1, \infty)$$

Now, we can determine the range of the function based on its domain as follows:

$$fog(x) = x \xrightarrow{\quad D_{fog} = [1, \infty) \quad} R_{fog} = [1, \infty)$$

Choice (2) is the answer.

8.42. Based on the information given in the problem, we have:

$$f(x) = \sqrt{x^2 - 2x + 3}$$

The problem can be solved as follows:

$$\Rightarrow f(x) = \sqrt{(x-1)^2 + 2}$$

As we know:

$$(x - 1)^2 \geq 0 \xrightarrow{+2} (x-1)^2 + 2 \geq 2 \xrightarrow{\sqrt{\quad}} \sqrt{(x-1)^2 + 2} \geq \sqrt{2} \Rightarrow f(x) \geq \sqrt{2} \Rightarrow R_{f(x)} = \left[\sqrt{2}, \infty\right)$$

Choice (1) is the answer.

8.43. Based on the information given in the problem, we have:

$$f(x) = |x - 2| \Rightarrow D_f = \mathbb{R}$$

Based on the definition, two functions are equivalent if they are equal and their domains are the same.

Choice (1):

$$g_1(x) = \left|\frac{x^2 - 3x + 2}{x - 1}\right| = \left|\frac{(x-1)(x-2)}{x-1}\right| = |x - 2|$$

$$\Rightarrow D_{g_1} = \mathbb{R} - \{1\}$$

Therefore, the functions are not equivalent, since their domains are different. Note that $x = 1$ makes the denominator zero, thus it is not in the domain.

Choice (2):

$$g_2(x) = \left|\frac{x^2 - 4}{x + 2}\right| = \left|\frac{(x - 2)(x + 2)}{x + 2}\right| = |x - 2|$$

$$\Rightarrow D_{g_2} = \mathbb{R} - \{-2\}$$

Therefore, the functions are not equivalent because their domains are not the same. Note that $x = -2$ makes the denominator zero, thus it must be excluded from the domain.

Choice (3):

$$g_3(x) = \frac{(x - 2)^2}{|x - 2|} = \frac{|x - 2|^2}{|x - 2|} = |x - 2|$$

$$\Rightarrow D_{g_3} = \mathbb{R} - \{-2\}$$

Therefore, the functions are not equivalent because their domains are not the same. Note that $x = -2$ makes the denominator 0, thus it must be excluded from the domain.

Choice (4):

$$g_4(x) = \frac{|6x - 12|}{6} = \frac{6|x - 2|}{6} = |x - 2|$$

$$\Rightarrow D_{g_4} = \mathbb{R}$$

Therefore, the functions are equivalent because their functions and domains are the same. Choice (4) is the answer.

Reference

1. Rahmani-Andebili, M. (2021). Precalculus – Practice Problems, Methods, and Solutions, Springer Nature, 2021.

Abstract

In this chapter, the basic and advanced problems of factorization of polynomials are presented. To help students study the chapter in the most efficient way, the problems are categorized into different levels based on their difficulty (easy, normal, and hard) and calculation amounts (small, normal, and large). Moreover, the problems are ordered from the easiest, with the smallest computations, to the most difficult, with the largest calculations.

9.1. Which one of the following terms is not a factor of $x^3 - 7x^2 + 6x$ [1]?

Difficulty level   ● Easy   ○ Normal   ○ Hard
Calculation amount   ● Small   ○ Normal   ○ Large

1) x
2) $x - 1$
3) $x - 6$
4) $x + 6$

Exercise: Which one of the following terms is not a factor of $x^3 - x^2 - 12x$?
1) x
2) $x - 1$
3) $x - 4$
4) $x + 3$

Final answer: Choice (2).

9.2. Calculate the value of the following term.

$$\frac{x^3 + 8}{x^2 - 2x + 4}$$

Difficulty level   ● Easy   ○ Normal   ○ Hard
Calculation amount   ● Small   ○ Normal   ○ Large

1) $x + 2$
2) $x - 2$
3) $x + 1$
4) $x - 1$

Exercise: Calculate the value of the term below.

$$\frac{x^3 - 27}{x - 3}$$

1) $x + 3$
2) $x - 3$
3) $x^2 - 3x + 9$
4) $x^2 + 3x + 9$

Final answer: Choice (4).

9.3. Which one of the choices is a factor of $x^3 + y^3$?

Difficulty level   ● Easy   ○ Normal   ○ Hard
Calculation amount   ● Small   ○ Normal   ○ Large

1) $x - y$
2) $x^2 + xy + y^2$
3) $x^2 - xy + y^2$
4) $x + y + xy$

9.4. Which one of the choices is a factor of $x^3 - y^3$?

Difficulty level   ● Easy   ○ Normal   ○ Hard
Calculation amount   ● Small   ○ Normal   ○ Large

1) $x^2 + xy + y^2$
2) $x^2 - xy + y^2$
3) $x + y$
4) $x + y + xy$

9.5. Which one of the choices is not a factor of $x^3 + 2x^2 - 3x$?

Difficulty level   ● Easy   ○ Normal   ○ Hard
Calculation amount   ● Small   ○ Normal   ○ Large

1) $x + 1$
2) x
3) $x - 1$
4) $x + 3$

Exercise: Which one of the following terms is not a factor of $x^3 + x^2 - 12x$?

1) x
2) $x - 3$
3) $x - 4$
4) $x + 4$

Final answer: Choice (3).

9.6. Which one of the choices is a factor of $6x^3 + 7x^2 + 2x$?

Difficulty level   ● Easy   ○ Normal   ○ Hard
Calculation amount   ● Small   ○ Normal   ○ Large

1) $3x + 2$
2) $2x - 1$
3) $3x^2 - 2x$
4) $2x^2 - x$

> ***Exercise:*** Which one of the choices is a factor of $2x^3 + 5x^2 + 3x$?
> 1) $2x + 3$
> 2) $x - 1$
> 3) $2x + 5$
> 4) $2x^2 + 5x$
>
> *Final answer*: Choice (1).

9.7. Which one of the choices is a factor of $x^3 + x - 10$?

Difficulty level   ○ Easy   ● Normal   ○ Hard
Calculation amount   ● Small   ○ Normal   ○ Large

1) $x^2 - 2x + 5$
2) $x^2 + 2x + 5$
3) $x + 2$
4) $x^2 - 1$

> ***Exercise:*** Which one of the choices is a factor of $x^3 + x + 30$?
> 1) $x - 3$
> 2) $x^2 + x + 10$
> 3) $x^2 + 3x + 10$
> 4) $x^2 - 3x + 10$
>
> *Final answer*: Choice (4).

9.8. Which one of the choices is a factor of $3a^2 + 2ab - b^2$?

Difficulty level   ○ Easy   ● Normal   ○ Hard
Calculation amount   ● Small   ○ Normal   ○ Large

1) $a - 2b$
2) $3a - b$
3) $a - b$
4) $2a - b$

9.9. Which one of the choices is not a factor of $x^5 - x^4 - 4x + 4$?

Difficulty level   ○ Easy   ● Normal   ○ Hard
Calculation amount   ● Small   ○ Normal   ○ Large

1) $x^2 - 2$
2) $x + 1$
3) $x^2 + 2$
4) $x - 1$

Exercise: Which one of the choices is not a factor of $x^5 - x^4 - 16x + 16$?

1) $x - 2$
2) $x + 2$
3) $x^2 + 4$
4) $x + 1$

Final answer: Choice (4).

9.10. Calculate the value of $x^2 + \frac{1}{x^2}$ if $x + \frac{1}{x} = 2$.

Difficulty level   ○ Easy   ● Normal   ○ Hard
Calculation amount   ● Small   ○ Normal   ○ Large

1) 8
2) 6
3) 4
4) 2

9.11. Calculate the value of the following term.

$$\frac{(x+1)^3 - 3x(x+1)}{x^3 + 1}$$

Difficulty level   ○ Easy   ● Normal   ○ Hard
Calculation amount   ● Small   ○ Normal   ○ Large

1) $\frac{x^3 + 1 - 3x^2 - 3x}{x^3 + 1}$
2) 1
3) $\frac{x^3 - 3x}{x^3 + 1}$
4) $\frac{x^3 - 3x^2}{x^3 + 1}$

9.12. Calculate the value of $(50.01)^2 - (49.99)^2$.

Difficulty level   ○ Easy   ● Normal   ○ Hard
Calculation amount   ● Small   ○ Normal   ○ Large

1) 0.2
2) 0.02
3) 2
4) 0

Exercise: Calculate the value of $(1.9)^2 - (1.1)^2$.

1) 0.24
2) 0.024
3) 2.4
4) 24

Final answer: Choice (3).

9.13. Which one of the following terms is not a factor of $x^4 + 2x^3 - x - 2$?

Difficulty level ○ Easy ● Normal ○ Hard
Calculation amount ● Small ○ Normal ○ Large
1) $x^2 - x + 1$
2) $x^2 + x + 1$
3) $x + 2$
4) $x - 1$

9.14. Which one of the following terms shows the correct factorization of $x^4 + x^2 - 2$?

Difficulty level ○ Easy ● Normal ○ Hard
Calculation amount ● Small ○ Normal ○ Large
1) $(x^2 - 2)(x^2 + 1)$
2) $(x^2 + 2)(x^2 + 1)$
3) $(x^2 + 2)(x + 1)(x - 1)$
4) $(x^2 - 2)(x + 1)(x - 1)$

9.15. Calculate the value of the term below.

$$\frac{x^3 - 2x^{\frac{3}{2}} + 1}{\left(x^{\frac{3}{2}} - 1\right)^2}$$

Difficulty level ○ Easy ● Normal ○ Hard
Calculation amount ● Small ○ Normal ○ Large
1) 2
2) $x - 2$
3) 1
4) x

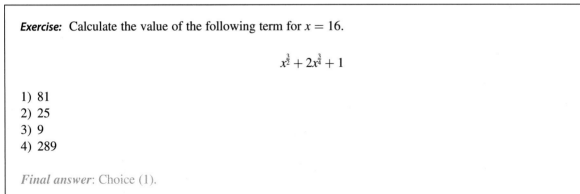

Exercise: Calculate the value of the following term for $x = 16$.

$$x^{\frac{3}{2}} + 2x^{\frac{3}{4}} + 1$$

1) 81
2) 25
3) 9
4) 289

Final answer: Choice (1).

9.16. Calculate the value of $a^3 - b^3$, if $a - b = 1$ and $a^2 + b^2 = 5$.
Difficulty level ○ Easy ● Normal ○ Hard
Calculation amount ○ Small ● Normal ○ Large
1) 10
2) 9
3) 7
4) 6

9.17. Calculate the value of the following term.

$$\frac{(a+b)^2 + (b+c)^2 + (a+c)^2 - (a+b+c)^2}{a^2 + b^2 + c^2}$$

Difficulty level   ○ Easy   ● Normal   ○ Hard
Calculation amount   ○ Small   ● Normal   ○ Large

1) 1
2) $\dfrac{ab + bc + ac}{a^2 + b^2 + c^2}$
3) $\dfrac{(a+b+c)^2}{a^2 + b^2 + c^2}$
4) $\dfrac{2(ab + bc + ac)}{a^2 + b^2 + c^2}$

Exercise: Calculate the value of the term below for $a = \sqrt{3}$, $b = -\sqrt{3}$, and $c = 2$.

$$a^2 + b^2 + c^2 + 2ab + 2bc + 2ac$$

1) 1
2) 2
3) 3
4) 4

Final answer: Choice (4).

9.18. Calculate the value of $x^3 + y^3$, if $xy = 5$ and $x + y = 7$.
Difficulty level   ○ Easy   ● Normal   ○ Hard
Calculation amount   ○ Small   ● Normal   ○ Large
1) 216
2) 238
3) 244
4) 264

9.19. Calculate the value of the term below.

$$\frac{2x^4 + x^3 - 4x - 2}{x^3 - 2}$$

Difficulty level   ○ Easy   ● Normal   ○ Hard
Calculation amount   ○ Small   ● Normal   ○ Large
1) $2x$
2) $2x - 1$
3) $2x + 1$
4) $x + 1$

Exercise: Which one of the following terms is not a factor of $2x^4 + x^3 + 16x + 8$?

1) $x + 2$
2) $2x + 1$
3) $x^2 + 4x + 4$
4) $x^2 - 4x + 4$

Final answer: Choice (3).

9.20. Calculate the value of the term below.

$$\frac{x^4 + x^2 + 1}{x^2 + x + 1}$$

Difficulty level   ○ Easy   ● Normal   ○ Hard
Calculation amount   ○ Small   ● Normal   ○ Large

1) $x^2 + x + 1$
2) $x^2 + x$
3) $x^2 - x + 1$
4) $x^2 + x - 1$

Exercise: Which one of the following terms is a factor of $x^4 + x^2 + 1$?

1) $x^2 - x + 1$
2) $x^2 - x - 1$
3) $x^2 + x - 1$
4) $x^2 + 1$

Final answer: Choice (1).

9.21. Calculate the value of the following term.

$$\frac{x^6 + 4x^2 + 5}{x^2 + 1}$$

Difficulty level   ○ Easy   ○ Normal   ● Hard
Calculation amount   ● Small   ○ Normal   ○ Large

1) $x^4 + x^2 + 5$
2) $x^4 - x^2 + 5$
3) $x^4 - 2x^2 + 5$
4) $x^4 + 2x^2 + 5$

9.22. Calculate the value of the following term.

$$\frac{x^{\frac{9}{2}} - 1}{x^3 + x^{\frac{3}{2}} + 1}$$

Difficulty level   ○ Easy   ○ Normal   ● Hard
Calculation amount   ● Small   ○ Normal   ○ Large

1) $x^{\frac{3}{2}} + 1$

2) $x^{\frac{3}{2}} - 1$

3) $x^{\frac{3}{2}} + 2$

4) $x^{\frac{3}{2}} - 2$

9.23. Calculate the value of $x^2 + y^2$, if $x^4 + y^4 + 4 = 2(2x^2 + 2y^2 - x^2 y^2)$.

Difficulty level   ○ Easy   ○ Normal   ● Hard

Calculation amount   ○ Small   ● Normal   ○ Large

1) 2

2) 1

3) 0

4) 8

Exercise: Calculate the value of the term below for $x = \sqrt{2}$ and $y = 1$.

$$x^4 + y^4 + 4 - 2\left(2x^2 + 2y^2 - x^2 y^2\right)$$

1) 0

2) 1

3) 2

4) 3

Final answer: Choice (2).

9.24. Calculate the value of the following term.

$$(2 + 1)\left(2^2 + 1\right)\left(2^4 + 1\right)\ldots\left(2^{64} + 1\right)$$

Difficulty level   ○ Easy   ○ Normal   ● Hard

Calculation amount   ○ Small   ● Normal   ○ Large

1) $2^{256} + 1$

2) $2^{128} + 1$

3) $2^{128} - 1$

4) $2^{256} - 1$

Exercise: Calculate the value of the following term.

$$(3 + 1)(9 + 1)(81 + 1)\ldots\left(3^{32} + 1\right)$$

1) $3^{128} + 1$

2) $3^{128} - 1$

3) $3^{64} + 1$

4) $3^{64} - 1$

Final answer: Choice (4).

Reference

1. Rahmani-Andebili, M. (2021). Precalculus – Practice Problems, Methods, and Solutions, Springer Nature, 2021.

Abstract

In this chapter, the problems of the ninth chapter are fully solved, in detail, step-by-step, and with different methods.

10.1. The problem can be solved as follows [1]:

$$x^3 - 7x^2 + 6x = x(x^2 - 7x + 6) = x(x-1)(x-6)$$

As can be seen, $x + 6$ is not a factor of the expression. Choice (4) is the answer.

10.2. Based on the rule of sum of two cubes, we know that:

$$a^3 + b^3 = (a + b)(a^2 - ab + b^2)$$

Therefore:

$$\frac{x^3 + 8}{x^2 - 2x + 4} = \frac{(x+2)(x^2 - 2x + 4)}{x^2 - 2x + 4} = x + 2$$

Choice (1) is the answer.

10.3. Based on the rule of sum of two cubes, we know that:

$$x^3 + y^3 = (x + y)(x^2 - xy + y^2)$$

Choice (3) is the answer.

10.4. Based on the rule of difference of two cubes, we know that:

$$x^3 - y^3 = (x - y)(x^2 + xy + y^2)$$

Choice (1) is the answer.

10.5. The problem can be solved as follows:

$$x^3 + 2x^2 - 3x = x(x^2 + 2x - 3) = x(x + 3)(x - 1)$$

Choice (1) is the answer.

10.6. The problem can be solved as follows:

$$6x^3 + 7x^2 + 2x = x\left(6x^2 + 7x + 2\right) = x(3x + 2)(2x + 1)$$

Choice (1) is the answer.

10.7. Based on the rule of difference of two cubes, we know that:

$$x^3 - y^3 = (x - y)\left(x^2 + xy + y^2\right)$$

Therefore:

$$x^3 + x - 10 = \left(x^3 - 8\right) + (x - 2) = (x - 2)\left(x^2 + 2x + 4\right) + (x - 2) = (x - 2)\left(x^2 + 2x + 5\right)$$

Choice (2) is the answer.

10.8. Based on the rule of difference of two squares, we know that:

$$(a + b)(a - b) = a^2 - b^2$$

The problem can be solved as follows:

$$3a^2 + 2ab - b^2 = 2a^2 + 2ab + a^2 - b^2 = 2a(a + b) + (a + b)(a - b) = (a + b)(2a + a - b) = (a + b)(3a - b)$$

Choice (2) is the answer.

10.9. From the rule of difference of two squares, we know that:

$$(a + b)(a - b) = a^2 - b^2$$

The problem can be solved as follows:

$$x^5 - x^4 - 4x + 4 = x^5 - 4x - \left(x^4 - 4\right) = x\left(x^4 - 4\right) - \left(x^4 - 4\right) = \left(x^4 - 4\right)(x - 1) = \left(x^2 + 2\right)\left(x^2 - 2\right)(x - 1)$$

Choice (2) is the answer.

10.10. Based on the information given in the problem, we have:

$$x + \frac{1}{x} = 2 \tag{1}$$

The problem can be solved as follows:

$$x^2 + \frac{1}{x^2} = x^2 + \frac{1}{x^2} + 2 - 2 = \left(x + \frac{1}{x}\right)^2 - 2 \tag{2}$$

Solving (1) and (2):

$$x^2 + \frac{1}{x^2} = (2)^2 - 2 = 2$$

Choice (4) is the answer.

10.11. As we know:

$$(a+b)^3 = a^3 + 3a^2b + 3ab^2 + b^3$$

Therefore:

$$\frac{(x+1)^3 - 3x(x+1)}{x^3+1} = \frac{x^3 + 3x^2 + 3x + 1 - 3x^2 - 3x}{x^3+1} = \frac{x^3+1}{x^3+1} = 1$$

Choice (2) is the answer.

10.12. Based on the rule of difference of two squares, we know that:

$$(a+b)(a-b) = a^2 - b^2$$

The problem can be solved as follows:

$$(50.01)^2 - (49.99)^2 = (50.01 + 49.99)(50.01 - 49.99) = (100)(0.02) = 2$$

Choice (3) is the answer.

10.13. Based on the rule of difference of two cubes, we know that:

$$a^3 - b^3 = (a-b)(a^2 + ab + b^2)$$

The problem can be solved as follows:

$$x^4 + 2x^3 - x - 2 = x^3(x+2) - (x+2) = (x+2)(x^3-1) = (x+2)(x-1)(x^2+x+1)$$

Choice (1) is the answer.

10.14. Based on the rule of difference of two squares, we know that:

$$(a+b)(a-b) = a^2 - b^2$$

The problem can be solved as follows:

$$x^4 + x^2 - 2 = x^4 - 1 + x^2 - 1 = (x^2-1)(x^2+1) + x^2 - 1 = (x^2-1)(x^2+2) = (x-1)(x+1)(x^2+2)$$

Choice (3) is the answer.

10.15. Based on the perfect square binomial formula, we have:

$$a^2 + b^2 - 2ab = (a-b)^2$$

The problem can be solved as follows:

$$\frac{x^3 - 2x^{\frac{3}{2}} + 1}{\left(x^{\frac{3}{2}} - 1\right)^2} = \frac{\left(x^{\frac{3}{2}}\right)^2 - 2x^{\frac{3}{2}} + 1}{\left(x^{\frac{3}{2}} - 1\right)^2} = \frac{\left(x^{\frac{3}{2}} - 1\right)^2}{\left(x^{\frac{3}{2}} - 1\right)^2} = 1$$

Choice (3) is the answer.

10.16. From the rule of difference of two cubes, we know that:

$$a^3 - b^3 = (a - b)(a^2 + b^2 + ab) \tag{1}$$

Based on the information given in the problem, we have:

$$a - b = 1 \tag{2}$$

$$a^2 + b^2 = 5 \tag{3}$$

From (2), we have:

$$a - b = 1 \xrightarrow{(\ )^2} (a - b)^2 = 1 \Rightarrow a^2 + b^2 - 2ab = 1 \tag{4}$$

Solving (3) and (4):

$$5 - 2ab = 1 \Rightarrow ab = 2 \tag{5}$$

Solving (1), (2), (3), and (5):

$$a^3 - b^3 = 1 \times (5 + 2) = 7$$

Choice (3) is the answer.

10.17. As we know:

$$(a + b + c)^2 = a^2 + b^2 + c^2 + 2ab + 2bc + 2ac$$

Therefore:

$$\frac{(a + b)^2 + (b + c)^2 + (a + c)^2 - (a + b + c)^2}{a^2 + b^2 + c^2}$$
$$= \frac{a^2 + b^2 + 2ab + b^2 + c^2 + 2bc + a^2 + c^2 + 2ac - (a^2 + b^2 + c^2 + 2ab + 2bc + 2ac)}{a^2 + b^2 + c^2}$$
$$= \frac{a^2 + b^2 + c^2}{a^2 + b^2 + c^2} = 1$$

Choice (1) is the answer.

10.18. Based on the information given in the problem, we have:

$$xy = 5$$

$$x + y = 7$$

Based on the rule of sum of two cubes, we know that:

$$x^3 + y^3 = (x + y)(x^2 - xy + y^2)$$

Therefore:

$$x^3 + y^3 = (x+y)(x^2 + 2xy + y^2 - 3xy) = (x+y)\left((x+y)^2 - 3xy\right) = 7 \times (7^2 - 3 \times 5) = 7 \times 34 = 238$$

Choice (2) is the answer.

10.19. The problem can be solved as follows:

$$\frac{2x^4 + x^3 - 4x - 2}{x^3 - 2} = \frac{(2x^4 - 4x) + (x^3 - 2)}{x^3 - 2} = \frac{2x(x^3 - 2) + (x^3 - 2)}{x^3 - 2} = \frac{(x^3 - 2)(2x + 1)}{x^3 - 2} = 2x + 1$$

Choice (3) is the answer.

10.20. Based on the rule of difference of two squares and perfect square binomial formula, we know that:

$$(a + b)(a - b) = a^2 - b^2$$

$$a^2 + b^2 + 2ab = (a + b)^2$$

The problem can be solved as follows:

$$\frac{x^4 + x^2 + 1}{x^2 + x + 1} = \frac{x^4 + x^2 + 1 + x^2 - x^2}{x^2 + x + 1} = \frac{x^4 + 2x^2 + 1 - x^2}{x^2 + x + 1} = \frac{(x^2 + 1)^2 - x^2}{x^2 + x + 1}$$
$$= \frac{(x^2 + 1 + x)(x^2 + 1 - x)}{x^2 + x + 1} = x^2 - x + 1$$

Choice (3) is the answer.

10.21. Based on the information given in the problem, the answer of the following term is requested.

$$\frac{x^6 + 4x^2 + 5}{x^2 + 1}$$

The problem can be solved as follows:

$$
\begin{array}{r|l}
x^6 + 4x^2 + 5 & x^2 + 1 \\
\underline{x^6 + x^4} & x^4 - x^2 + 5 \\
-x^4 + 4x^2 & \\
\underline{-x^4 - x^2} & \\
5x^2 + 5 & \\
\underline{5x^2 + 5} & \\
0 &
\end{array}
$$

Choice (2) is the answer.

10.22. Based on the rule of difference of two cubes, we know that:

$$a^3 - b^3 = (a-b)(a^2 + ab + b^2)$$

Therefore:

$$\frac{x^{\frac{9}{2}} - 1}{x^3 + x^{\frac{3}{2}} + 1} = \frac{\left(x^{\frac{3}{2}} - 1\right)\left(\left(x^{\frac{3}{2}}\right)^2 + x^{\frac{3}{2}} + 1\right)}{\left(x^{\frac{3}{2}}\right)^2 + x^{\frac{3}{2}} + 1} = x^{\frac{3}{2}} - 1$$

Choice (2) is the answer.

10.23. As we know:

$$(a + b + c)^2 = a^2 + b^2 + c^2 + 2(ab + bc + ac) \tag{1}$$

Therefore:

$$x^4 + y^4 + 4 = 2(2x^2 + 2y^2 - x^2 y^2) \Rightarrow x^4 + y^4 + 4 - 2(2x^2 + 2y^2 - x^2 y^2) = 0$$

$$\Rightarrow \left(-x^2\right)^2 + \left(-y^2\right)^2 + 2^2 + 2\left(2\left(-x^2\right) + 2\left(-y^2\right) + \left(-x^2\right)\left(-y^2\right)\right) = 0 \tag{2}$$

Solving (1) and (2):

$$\left(\left(-x^2\right) + \left(-y^2\right) + 2\right)^2 = 0 \Rightarrow -x^2 - y^2 + 2 = 0 \Rightarrow x^2 + y^2 = 2$$

Choice (1) is the answer.

10.24. Based on the rule of difference of two squares, we know that:

$$(a + b)(a - b) = a^2 - b^2$$

Now, the problem can be solved as follows:

$$(2 + 1)\left(2^2 + 1\right)\left(2^4 + 1\right) \ldots \left(2^{64} + 1\right)$$

$$\xrightarrow{\times (2 - 1)} (2 - 1)(2 + 1)\left(2^2 + 1\right)\left(2^4 + 1\right) \ldots \left(2^{64} + 1\right) = \left(2^2 - 1\right)\left(2^2 + 1\right)\left(2^4 + 1\right) \ldots \left(2^{64} + 1\right)$$

$$= \left(2^4 - 1\right)\left(2^4 + 1\right) \ldots \left(2^{64} + 1\right) = \left(2^8 - 1\right) \ldots \left(2^{64} + 1\right) = \ldots = \left(2^{64} - 1\right)\left(2^{64} + 1\right) = 2^{128} - 1$$

Choice (3) is the answer.

Reference

1. Rahmani-Andebili, M. (2021). Precalculus – Practice Problems, Methods, and Solutions, Springer Nature, 2021.

Problems: Trigonometric and Inverse Trigonometric Functions

Abstract

In this chapter, the basic and advanced problems of trigonometric and inverse trigonometric functions are presented. To help students study the chapter in the most efficient way, the problems are categorized into different levels based on their difficulty (easy, normal, and hard) and calculation amounts (small, normal, and large). Moreover, the problems are ordered from the easiest, with the smallest computations, to the most difficult, with the largest calculations.

11.1. Calculate the value of $2\cos(x)$, if $\sin(x) = \frac{1}{2}$ and the end of arc x is in the second quadrant [1].

Difficulty level   ● Easy   ○ Normal   ○ Hard
Calculation amount   ● Small   ○ Normal   ○ Large

1) $-\sqrt{3}$
2) $-\sqrt{2}$
3) $\sqrt{2}$
4) $\sqrt{3}$

Exercise: Calculate the value of $\sin(x)$, if $\cos(x) = \frac{1}{2}$ and the end of arc x is in the fourth quadrant.

Difficulty level   ● Easy   ○ Normal   ○ Hard
Calculation amount   ● Small   ○ Normal   ○ Large

1) $\frac{\sqrt{3}}{2}$
2) $-\frac{\sqrt{3}}{2}$
3) $-\frac{1}{2}$
4) $\frac{1}{2}$

Final answer: Choice (2).

11.2. If $\cos(\alpha)(1 + \tan^2(\alpha)) > 0$, determine the quadrant where the end of arc α is located.

Difficulty level   ● Easy   ○ Normal   ○ Hard
Calculation amount   ● Small   ○ Normal   ○ Large

1) First
2) Second
3) First or third
4) First or fourth

Exercise: If $\sin(\alpha)(1 + \sin^2(\alpha)) > 0$, determine the quadrant where the end of arc α is located.

Difficulty level   ● Easy   ○ Normal   ○ Hard
Calculation amount   ● Small   ○ Normal   ○ Large

1) First
2) Second
3) First or second
4) First or fourth

Final answer: Choice (3).

11.3. Calculate the value of $\cos(10°) + \cos(190°)$.

Difficulty level   ● Easy   ○ Normal   ○ Hard
Calculation amount   ● Small   ○ Normal   ○ Large

1) 0
2) $2\cos(10°)$
3) $2\sin(10°)$
4) $\cos(200°)$

Exercise: Calculate the value of $\sin(20°) + \sin(200°)$.

1) $\sin(220°)$
2) $2\sin(20°)$
3) $2\cos(20°)$
4) 0

Final answer: Choice (4).

11.4. Calculate the value of $\sin(200°) + \sin(10°) + \sin(20°) + \sin(190°)$.

Difficulty level   ● Easy   ○ Normal   ○ Hard
Calculation amount   ● Small   ○ Normal   ○ Large

1) $2\sin(10°)$
2) $2\cos(10°)$
3) 0
4) $-\sin(10°)$

Exercise: Calculate the value of $\cos(5°) + \cos(-5°)$.

1) $\cos(10°)$
2) $2\cos(5°)$
3) $-\cos(10°)$
4) 0

Final answer: Choice (2).

11.5. Calculate the value of $\sin(1860°) + \cos(1860°)$.

1) $\frac{1+\sqrt{3}}{2}$

2) $\frac{\sqrt{3}-1}{2}$

3) $\frac{1-\sqrt{3}}{2}$

4) 0

Exercise: Calculate the value of $\sin(3630°) + \cos(3630°)$.

1) $\frac{1+\sqrt{3}}{2}$

2) $\frac{\sqrt{3}-1}{2}$

3) $\frac{1-\sqrt{3}}{2}$

4) 0

Final answer: Choice (1).

11.6. Calculate the value of $\tan(135°)\sin^2(60°)$.

1) $-\frac{3}{4}$

2) $-\frac{1}{2}$

3) $\frac{1}{4}$

4) $\frac{3}{4}$

Exercise: Calculate the value of $\cot(135°)\cos^2(60°)$.

1) $\frac{1}{4}$

2) $-\frac{1}{2}$

3) $-\frac{1}{4}$

4) $\frac{\sqrt{3}}{2}$

Final answer: Choice (3).

11.7. Calculate the value of $\cos^2\left(\frac{5\pi}{4}\right) + \sin\left(\frac{7\pi}{6}\right)$.

1) 1

2) $\frac{1}{2}$

3) $\frac{1}{4}$

4) 0

Exercise: Calculate the value of $\sin \frac{3\pi}{4} + \cos \frac{3\pi}{4}$.

1) $\sqrt{2}$

2) $\frac{\sqrt{2}}{2}$

3) 0

4) $2\sqrt{2}$

Final answer: Choice (3).

11.8. Calculate the value of $(\sin(60°) - \sin(45°))(\cos(30°) + \cos(45°))$.

Difficulty level   ● Easy   ○ Normal   ○ Hard

Calculation amount   ● Small   ○ Normal   ○ Large

1) $\frac{3}{4}$

2) $\frac{1}{4}$

3) $\frac{1}{2}$

4) 1

11.9. If $\cos(\theta) = -\frac{2}{3}$ and $\tan(\theta) \cos(\theta) > 0$, which quadrant is the location of the end of arc θ?

Difficulty level   ● Easy   ○ Normal   ○ Hard

Calculation amount   ● Small   ○ Normal   ○ Large

1) First

2) Second

3) Third

4) Fourth

Exercise: If $\sin(\theta) < 0$ and $\sin(\theta) \cot(\theta) > 0$, which quadrant is the location of the end of arc θ?

1) First

2) Second

3) Third

4) Fourth

Final answer: Choice (4).

11.10. In Fig. 11.1, in triangle ABC, determine the value of $\sin(\theta)$.

Difficulty level   ● Easy   ○ Normal   ○ Hard

Calculation amount   ● Small   ○ Normal   ○ Large

1) $\frac{3}{5}$

2) $\frac{3}{4}$

3) $\frac{4}{5}$

4) $\frac{5}{4}$

Exercise: In the triangle shown in Fig. 11.1, calculate the value of $\cos(\theta)$.

Difficulty level   ● Easy   ○ Normal   ○ Hard

Calculation amount   ● Small   ○ Normal   ○ Large

1) $\dfrac{3}{5}$

2) $\dfrac{3}{4}$

3) $\dfrac{4}{5}$

4) $\dfrac{5}{4}$

Final answer: Choice (3).

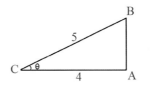

Figure 11.1 The graph of problem 11.10

11.11. In Fig. 11.2, determine the value of $\tan(\alpha) + \tan(\beta)$.

Difficulty level   ● Easy   ○ Normal   ○ Hard

Calculation amount   ● Small   ○ Normal   ○ Large

1) $\dfrac{5}{6}$

2) $\dfrac{4}{5}$

3) $\dfrac{6}{5}$

4) $\dfrac{5}{4}$

Exercise: In Fig. 11.2, calculate the value of $\sin(\alpha) + \cos(\beta)$.

1) $\dfrac{\sqrt{10}+6\sqrt{5}}{10}$

2) $\dfrac{\sqrt{10}+4\sqrt{5}}{10}$

3) $\dfrac{\sqrt{10}+2\sqrt{5}}{10}$

4) $\dfrac{\sqrt{10}+\sqrt{5}}{10}$

Final answer: Choice (2).

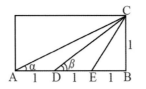

Figure 11.2 The graph of problem 11.11

11.12. Calculate the value of cos(60°) cos (30°) + sin (60°) sin (30°).

Difficulty level   ● Easy   ○ Normal   ○ Hard
Calculation amount   ● Small   ○ Normal   ○ Large
1) tan(30°)
2) cot(45°)
3) sin(60°)
4) cos(60°)

11.13. Which one of the statements below is correct?

Difficulty level   ○ Easy   ● Normal   ○ Hard
Calculation amount   ● Small   ○ Normal   ○ Large
1) sin(50°) < sin (40°)
2) cos(50°) < cos (40°)
3) tan(50°) < tan (40°)
4) cot(40°) < cot (50°)

Exercise: Which one of the statements below is incorrect?

1) sin(10°) < sin (20°)
2) cos(10°) > cos (20°)
3) tan(10°) < tan (20°)
4) cot(10°) < cot (20°)

Final answer: Choice (4).

11.14. Simplify and calculate the final answer for the term below.

$$\frac{\sin(60°)\cos(60°)\tan(60°)\csc(60°)\sec(60°)\cot(60°)}{\sin(40°)\cos(40°)\tan(40°)\csc(40°)\sec(40°)\cot(40°)}$$

Difficulty level   ○ Easy   ● Normal   ○ Hard
Calculation amount   ○ Small   ● Normal   ○ Large
1) 0
2) $\frac{1}{2}$
3) 1
4) 2

11.15. Calculate the value of sin(135°) + cos (45°) + tan (225°) + cot (315°).

Difficulty level   ○ Easy   ● Normal   ○ Hard
Calculation amount   ○ Small   ● Normal   ○ Large

1) $2 + \sqrt{2}$
2) $\sqrt{2}$
3) $\sqrt{2} - 2$
4) 2

11.16. Which one of the choices is equal to tan(10°)?

Difficulty level ○ Easy ● Normal ○ Hard
Calculation amount ○ Small ● Normal ○ Large

1) $\tan(-10°)$
2) $\cot(100°)$
3) $\tan(170°)$
4) $\tan(190°)$

Exercise: Which one of the choices is equal to sin(10°)?

Difficulty level ○ Easy ● Normal ○ Hard
Calculation amount ○ Small ● Normal ○ Large

1) $\sin(-10°)$
2) $\sin(170°)$
3) $\cos(-10°)$
4) $\cos(170°)$

Final answer: Choice (2).

11.17. Calculate the value of tan(θ), if $\sin(\theta) = -\frac{\sqrt{5}}{5}$ and the end of arc θ is in the third quadrant.

Difficulty level ○ Easy ● Normal ○ Hard
Calculation amount ○ Small ● Normal ○ Large

1) -2
2) $-\frac{1}{2}$
3) $\frac{1}{2}$
4) 2

Exercise: Calculate the value of tan(θ), if $\cos(\theta) = \frac{\sqrt{3}}{3}$ and the end of arc θ is in the fourth quadrant.

1) $\sqrt{2}$
2) $-\sqrt{2}$
3) $\frac{\sqrt{2}}{2}$
4) $-\frac{\sqrt{2}}{2}$

Final answer: Choice (2).

11.18. Which one of the choices is equivalent to the angle of $-27°$?

Difficulty level   ○ Easy   ● Normal   ○ Hard
Calculation amount   ○ Small   ● Normal   ○ Large
1) 127°
2) −127°
3) 333°
4) 53°

Exercise: Which one of the choices is not equivalent to the angle of 1°?

1) 361°
2) −1°
3) 721°
4) 1081°

Final answer: Choice (2).

11.19. Calculate the range of m if:

$$\cos(3x) = \frac{m-1}{2}, \quad -\frac{\pi}{9} < x < \frac{\pi}{9}$$

Difficulty level   ○ Easy   ○ Normal   ● Hard
Calculation amount   ● Small   ○ Normal   ○ Large
1) (1, 2]
2) (0, 2)
3) (2, 3]
4) (2, 3)

Exercise: Calculate the range of m if:

$$\cos(2x) = \frac{m+1}{2}, \quad -\frac{\pi}{12} < x < \frac{\pi}{12}$$

Difficulty level   ○ Easy   ○ Normal   ● Hard
Calculation amount   ● Small   ○ Normal   ○ Large
1) $\sqrt{3} < m < 2$
2) $0 < m < 1$
3) $\sqrt{3} < m < 1$
4) $\sqrt{3} - 1 < m < 1$

Final answer: Choice (4).

11.20. Calculate the range of m if:

$$\sin(x) = \frac{m}{2}, \quad \frac{\pi}{4} < x < \frac{3\pi}{5}$$

Difficulty level   ○ Easy   ○ Normal   ● Hard
Calculation amount   ● Small   ○ Normal   ○ Large
1) $m \leq \frac{\sqrt{2}}{2}$
2) $m > \frac{\sqrt{2}}{2}$
3) $\sqrt{2} < m \leq 2$
4) $1 \leq m < \sqrt{2}$

11.21. Calculate the range of m if:

$$\cos(2\alpha) = \frac{1}{1-m}, \quad \frac{\pi}{4} < \alpha < \frac{3\pi}{4}$$

Difficulty level   ○ Easy   ○ Normal   ● Hard
Calculation amount   ○ Small   ● Normal   ○ Large
1) $(-\infty, 2)$
2) $(1, \infty)$
3) $[2, \infty)$
4) $(-\infty, 1)$

11.22. Calculate the range of m if:

$$\sin(x) = \frac{3 - m^2}{3 + m^2}, \quad \frac{\pi}{3} < x < \frac{5\pi}{6}$$

Difficulty level   ○ Easy   ○ Normal   ● Hard
Calculation amount   ○ Small   ● Normal   ○ Large
1) $|m| < \sqrt{3}$
2) $|m| < \sqrt{2}$
3) $|m| < 1$
4) $|m| < \frac{1}{2}$

11.23. If $\sin(\alpha) \cos(\alpha) > 0$ and $\cos(\alpha) \cot(\alpha) < 0$, then determine the quadrant where the end of arc α is located.
Difficulty level   ○ Easy   ○ Normal   ● Hard
Calculation amount   ○ Small   ● Normal   ○ Large
1) First
2) Second
3) Third
4) Fourth

Exercise: If $\sin(\alpha) \cos(\alpha) < 0$ and $\sin(\alpha) \tan(\alpha) < 0$, then determine the quadrant where the end of arc α is located.
1) First
2) Second
3) Third
4) Second or fourth

Final answer: Choice (2).

11.24. **Which one of the inequalities is correct?**

Difficulty level   ○ Easy   ○ Normal   ● Hard
Calculation amount   ○ Small   ● Normal   ○ Large

1) $\sin(20°) > \sin(170°)$
2) $\cos(20°) < \cos(160°)$
3) $\sin(20°) < \sin(10°)$
4) $\cos(10°) < \cos(20°)$

Reference

1. Rahmani-Andebili, M. (2021). Precalculus – Practice Problems, Methods, and Solutions, Springer Nature, 2021.

Abstract

In this chapter, the problems of the 11th chapter are fully solved, in detail, step-by-step, and with different methods.

12.1. Based on the information given in the problem, we know that [1]:

$$\frac{\pi}{2} \leq x \leq \pi$$

$$\sin(x) = \frac{1}{2}$$

Moreover, from trigonometry, we know that:

$$\sin^2(x) + \cos^2(x) = 1$$

The problem can be solved as follows:

$$2\cos(x) = 2\sqrt{1 - \sin^2(x)} = 2\sqrt{1 - \left(\frac{1}{2}\right)^2} = 2\sqrt{\frac{3}{4}} = \pm\sqrt{3}$$

As can be noticed from Figure 12.1, $\sqrt{3}$ is not acceptable for $\cos(x)$, because the end of arc x is in the second quadrant $(\cos(x) < 0)$. Therefore:

$$2\cos(x) = -\sqrt{3}$$

Choice (1) is the answer.

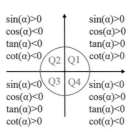

Figure 12.1 The graph of the solution of problem 12.1

12.2. Based on the information given in the problem, we know that:

$$\cos(\alpha)\left(1 + \tan^2(\alpha)\right) > 0 \xrightarrow{\; 1 + \tan^2(\alpha) > 0 \;} \cos(\alpha) > 0$$

Therefore, the end of arc α is in the first or fourth quadrant, as can be noticed in Figure 12.2. Choice (4) is the answer.

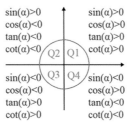

Figure 12.2 The graph of the solution of problem 12.2

12.3. From trigonometry, we know that for an acute angle of θ:

$$\cos(180^\circ + \theta) = -\cos(\theta)$$

Therefore:

$$\cos(10^\circ) + \cos(190^\circ) = \cos(10^\circ) + \cos(180^\circ + 10^\circ) = \cos(10^\circ) - \cos(10^\circ) = 0$$

Choice (1) is the answer.

12.4. From trigonometry, we know that for an acute angle of θ:

$$\sin(180^\circ + \theta) = -\sin(\theta)$$

Therefore:

$$\sin(200^\circ) + \sin(10^\circ) + \sin(20^\circ) + \sin(190^\circ)$$
$$= \sin(180^\circ + 20^\circ) + \sin(10^\circ) + \sin(20^\circ) + \sin(180^\circ + 10^\circ)$$
$$= -\sin(20^\circ) + \sin(10^\circ) + \sin(20^\circ) - \sin(10^\circ) = 0$$

Choice (3) is the answer.

12.5. From trigonometry, we know that for an acute angle of θ:

$$\sin(n \times 360° + \theta) = \sin(\theta), n \in \mathbb{Z}$$

$$\cos(n \times 360° + \theta) = \cos(\theta), n \in \mathbb{Z}$$

Moreover, we know that:

$$\sin(60°) = \frac{\sqrt{3}}{2}$$

$$\cos(60°) = \frac{1}{2}$$

Therefore:

$$\sin(1860°) + \cos(1860°) = \sin(5 \times 360° + 60°) + \cos(5 \times 360° + 60°)$$

$$= \sin(60°) + \cos(60°) = \frac{\sqrt{3}}{2} + \frac{1}{2} = \frac{\sqrt{3} + 1}{2}$$

Choice (1) is the answer.

12.6. From trigonometry, we know that for an acute angle of θ:

$$\tan(180° - \theta) = -\tan(\theta)$$

In addition, we know that:

$$\tan(45°) = 1$$

$$\sin(60°) = \frac{\sqrt{3}}{2}$$

Therefore:

$$\tan(135°) \sin^2(60°) = \tan(135°) \sin^2(60°) = \tan(180° - 45°) \sin^2(60°)$$

$$= -\tan(45°) \sin^2(60°) = -1 \times \left(\frac{\sqrt{3}}{2}\right)^2 = -\frac{3}{4}$$

Choice (1) is the answer.

12.7. From trigonometry, we know that for an acute angle of θ:

$$\cos(\pi + \theta) = -\cos(\theta)$$

$$\sin(\pi + \theta) = -\sin(\theta)$$

Moreover, we know that:

$$\cos\left(\frac{\pi}{4}\right) = \frac{\sqrt{2}}{2}$$

$$\sin\left(\frac{\pi}{6}\right) = \frac{1}{2}$$

Therefore:

$$\cos^2\left(\frac{5\pi}{4}\right) + \sin\left(\frac{7\pi}{6}\right) = \cos^2\left(\pi + \frac{\pi}{4}\right) + \sin\left(\pi + \frac{\pi}{6}\right) = \left(-\cos\left(\frac{\pi}{4}\right)\right)^2 - \sin\left(\frac{\pi}{6}\right) = \left(-\frac{\sqrt{2}}{2}\right)^2 - \frac{1}{2} = \frac{1}{2} - \frac{1}{2} = 0$$

Choice (4) is the answer.

12.8. From trigonometry, we know that:

$$\sin(60°) = \frac{\sqrt{3}}{2}$$

$$\sin(45°) = \frac{\sqrt{2}}{2}$$

$$\cos(30°) = \frac{\sqrt{3}}{2}$$

$$\cos(45°) = \frac{\sqrt{2}}{2}$$

In addition, based on the rule of difference of two squares, we know that:

$$(a + b)(a - b) = a^2 - b^2$$

The problem can be solved as follows:

$$(\sin(60°) - \sin(45°))(\cos(30°) + \cos(45°)) = \left(\frac{\sqrt{3}}{2} - \frac{\sqrt{2}}{2}\right)\left(\frac{\sqrt{3}}{2} + \frac{\sqrt{2}}{2}\right) = \left(\frac{\sqrt{3}}{2}\right)^2 - \left(\frac{\sqrt{2}}{2}\right)^2 = \frac{3}{4} - \frac{2}{4} = \frac{1}{4}$$

Choice (2) is the answer.

12.9. Based on the information given in the problem, we know that:

$$\cos(\theta) = -\frac{2}{3} \Rightarrow \cos(\theta) < 0 \tag{1}$$

$$\tan(\theta)\cos(\theta) > 0 \tag{2}$$

Solving (1) and (2):

$$\tan(\theta) < 0 \tag{3}$$

By considering Figure 12.3 as well as (1) and (3), we can conclude the location of the end of arc θ is in the second quadrant. Choice (2) is the answer.

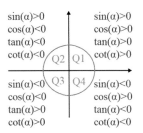

Figure 12.3 The graph of the solution of problem 12.9

12.10. From trigonometry, we know that:

$$\sin(\text{angle}) = \frac{\text{Side opposite to the angle}}{\text{Hypotenuse}}$$

Moreover, using the Pythagorean formula for the triangle shown in Figure 12.4, we can write:

$$BC^2 = AB^2 + AC^2 \Rightarrow AB = \sqrt{BC^2 - AC^2} = \sqrt{5^2 - 4^2} = 3$$

Therefore:

$$\sin(\theta) = \frac{AB}{BC} = \frac{3}{5}$$

Choice (1) is the answer.

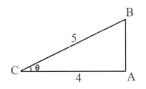

Figure 12.4 The graph of the solution of problem 12.10

12.11. From trigonometry, we know that:

$$\tan(\text{angle}) = \frac{\text{Side opposite to the angle}}{\text{Side adjacent to the angle}}$$

Therefore:

$$\tan(\alpha) + \tan(\beta) = \frac{CB}{AB} + \frac{CB}{DB} = \frac{1}{1+1+1} + \frac{1}{1+1} = \frac{1}{3} + \frac{1}{2} = \frac{5}{6}$$

Choice (1) is the answer.

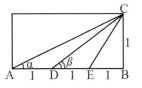

Figure 12.5 The graph of the solution of problem 12.11

12.12. From trigonometry, we know that:

$$\cos(60°) = \frac{1}{2}$$

$$\cos(30°) = \frac{\sqrt{3}}{2}$$

$$\sin(60°) = \frac{\sqrt{3}}{2}$$

$$\sin(30°) = \frac{1}{2}$$

Therefore:

$$\cos(60°)\cos(30°) + \sin(60°)\sin(30°) = \frac{1}{2} \times \frac{\sqrt{3}}{2} + \frac{\sqrt{3}}{2} \times \frac{1}{2} = \frac{\sqrt{3}}{4} + \frac{\sqrt{3}}{4} = \frac{\sqrt{3}}{2} = \sin(60°)$$

Choice (3) is the answer.

12.13. We know that for an acute angle, $\sin(\alpha)$ and $\tan(\alpha)$ have ascending trends. Therefore:

$$\sin(40°) < \sin(50°)$$

$$\tan(40°) < \tan(50°)$$

Moreover, for an acute angle, $\cos(\alpha)$ and $\cot(\alpha)$ are descending. Therefore:

$$\cos(50°) < \cos(40°)$$

$$\cot(50°) < \cot(40°)$$

Choice (2) is the answer.

12.14. From trigonometry, we know that:

$$\cot(\alpha) = \frac{1}{\tan(\alpha)}$$

$$\sec(\alpha) = \frac{1}{\cos(\alpha)}$$

$$\csc(\alpha) = \frac{1}{\sin(\alpha)}$$

The problem can be solved as follows:

$$\frac{\sin(60°)\cos(60°)\tan(60°)\csc(60°)\sec(60°)\cot(60°)}{\sin(40°)\cos(40°)\tan(40°)\csc(40°)\sec(40°)\cot(40°)}$$

$$= \frac{\sin(60°)\csc(60°)\cos(60°)\sec(60°)\tan(60°)\cot(60°)}{\sin(40°)\csc(40°)\cos(40°)\sec(40°)\tan(40°)\cot(40°)}$$

$$= \frac{\sin(60°)\dfrac{1}{\sin(60°)}\cos(60°)\dfrac{1}{\cos(60°)}\tan(60°)\dfrac{1}{\tan(60°)}}{\sin(40°)\dfrac{1}{\sin(40°)}\cos(40°)\dfrac{1}{\cos(40°)}\tan(40°)\dfrac{1}{\tan(40°)}} = \frac{1 \times 1 \times 1}{1 \times 1 \times 1} = 1$$

Choice (3) is the answer.

12.15. From trigonometry, we know that for an acute angle of θ:

$$\sin(180° - \theta) = \sin(\theta)$$

$$\tan(180° + \theta) = \tan(\theta)$$

$$\cot(360° - \theta) = -\cot(\theta)$$

In addition, we know that:

$$\sin(45°) = \frac{\sqrt{2}}{2}$$

$$\cos(45°) = \frac{\sqrt{2}}{2}$$

$$\tan(45°) = 1$$

$$\cot(45°) = 1$$

The problem can be solved as follows:

$$\sin(135°) + \cos(45°) + \tan(225°) + \cot(315°)$$
$$= \sin(180° - 45°) + \cos(45°) + \tan(180° + 45°) + \cot(360° - 45°)$$
$$= \sin(45°) + \cos(45°) + \tan(45°) - \cot(45°) = \frac{\sqrt{2}}{2} + \frac{\sqrt{2}}{2} + 1 - 1 = \sqrt{2}$$

Choice (2) is the answer.

12.16. From trigonometry, we know that for an acute angle of θ:

$$\tan(-\theta) = -\tan(\theta)$$

$$\cot(90° + \theta) = -\tan(\theta)$$

$$\tan(180° - \theta) = -\tan(\theta)$$

$$\tan(180° + \theta) = \tan(\theta)$$

Choice (1):

$$\tan(-10°) = -\tan(10°)$$

Choice (2):

$$\cot(100°) = \cot(90° + 10°) = -\tan(10°)$$

Choice (3):

$$\tan(170°) = \tan(180° - 10°) = -\tan(10°)$$

Choice (4):

$$\tan(190°) = \tan(180° + 10°) = \tan(10°)$$

Choice (4) is the answer.

12.17. Based on the information given in the problem, we know that the end of arc θ is in the third quadrant and:

$$\sin(\theta) = -\frac{\sqrt{5}}{5} \tag{1}$$

$$\sin^2(\theta) + \cos^2(\theta) = 1 \tag{2}$$

$$\tan(\theta) = \frac{\sin(\theta)}{\cos(\theta)} \tag{3}$$

The problem can be solved as follows:

Solving (1) and (2):

$$\cos(\theta) = \sqrt{1 - \sin^2(\theta)} = \sqrt{1 - \left(\frac{\sqrt{5}}{5}\right)^2} = \sqrt{1 - \frac{1}{5}} = \pm\frac{2}{\sqrt{5}} \tag{4}$$

In (4), the negative value is acceptable, because the end of arc θ is in the third quadrant (*See* Figure 12.6).

Solving (1), (3), and (4):

$$\tan(\theta) = \frac{\sin(\theta)}{\cos(\theta)} = \frac{-\frac{\sqrt{5}}{5}}{-\frac{2}{\sqrt{5}}} = \frac{1}{2}$$

Choice (3) is the answer.

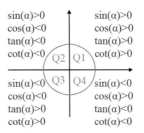

Figure 12.6 The graph of the solution of problem 12.17

12.18. As we know from trigonometry, in a unit circle, the relation below holds for any pair of angles:

$$\theta_2 - \theta_1 = n \times 360^\circ, n \in \mathbb{Z} \Rightarrow \theta_2 \equiv \theta_1$$

The problem can be solved as follows:

Choice (1):

$$127^\circ - (-27^\circ) = 154^\circ \Rightarrow 127^\circ \not\equiv -27^\circ$$

Choice (2):

$$-127^\circ - (-27^\circ) = 100^\circ \Rightarrow -127^\circ \not\equiv -27^\circ$$

Choice (3):

$$333^\circ - (-27^\circ) = 360^\circ \Rightarrow 333^\circ \equiv -27^\circ$$

Choice (4):

$$53^\circ - (-27^\circ) = 80^\circ \Rightarrow 53^\circ \not\equiv -27^\circ$$

Choice (3) is the answer.

12.19. Calculate the range of m if:

$$\cos(3x) = \frac{m-1}{2}, \quad -\frac{\pi}{9} < x < \frac{\pi}{9}$$

The problem can be solved as follows:

$$-\frac{\pi}{9} < x < \frac{\pi}{9} \overset{\times 3}{\Rightarrow} -\frac{\pi}{3} < 3x < \frac{\pi}{3} \tag{1}$$

Equation (1) and Figure 12.7 show the range of $3x$. Therefore:

$$\Rightarrow \frac{1}{2} < \cos(3x) \leq 1 \Rightarrow \frac{1}{2} < \frac{m-1}{2} \leq 1 \Rightarrow 1 < m-1 \leq 2 \Rightarrow 2 < m \leq 3$$

Choice (3) is the answer.

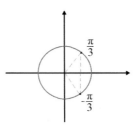

Figure 12.7 The graph of the solution of problem 12.19

12.20. Calculate the range of m if:

$$\sin(x) = \frac{m}{2} \tag{1}$$

$$\frac{\pi}{4} < x < \frac{3\pi}{5} \tag{2}$$

Equation (2) and Figure 12.8 show the range of x. Therefore:

$$\Rightarrow \frac{\sqrt{2}}{2} < \sin(x) \leq 1 \Rightarrow \frac{\sqrt{2}}{2} < \frac{m}{2} \leq 1 \Rightarrow \sqrt{2} < m \leq 2$$

Choice (3) is the answer.

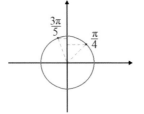

Figure 12.8 The graph of the solution of problem 12.20

12.21. Based on the information given in the problem, we have:

$$\cos(2\alpha) = \frac{1}{1-m}, \quad \frac{\pi}{4} < \alpha < \frac{3\pi}{4}$$

The problem can be solved as follows:

$$\frac{\pi}{4} < \alpha < \frac{3\pi}{4} \xrightarrow{\times 2} \frac{\pi}{2} < 2\alpha < \frac{3\pi}{2} \tag{1}$$

As can be seen in (1) and Figure 12.9, 2α is in the second and third quadrants. Therefore:

$$\Rightarrow -1 \leq \cos(2\alpha) < 0 \Rightarrow -1 \leq \frac{1}{1-m} < 0$$

$$\Rightarrow \begin{cases} \dfrac{1}{1-m} < 0 \Rightarrow 1-m < 0 \Rightarrow m > 1 \\ \dfrac{1}{1-m} \geq -1 \Rightarrow \dfrac{1}{1-m} + 1 \geq 0 \Rightarrow \dfrac{1+1-m}{1-m} \geq 0 \Rightarrow \dfrac{2-m}{1-m} \geq 0 \Rightarrow \dfrac{m-2}{m-1} \geq 0 \Rightarrow m < 1, m \geq 2 \end{cases} \overset{\cap}{\Rightarrow} m \geq 2$$

Choice (3) is the answer.

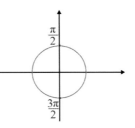

Figure 12.9 The graph of the solution of problem 12.21

12.22. Calculate the range of m if:

$$\sin(x) = \frac{3 - m^2}{3 + m^2}, \quad \frac{\pi}{3} < x < \frac{5\pi}{6}$$

The problem can be solved as follows:

$$\frac{\pi}{3} < x < \frac{5\pi}{6} \tag{1}$$

Equation (1) and Figure 12.10 show the range of x. Therefore:

$$\Rightarrow \frac{1}{2} \leq \sin(x) < 1 \Rightarrow \frac{1}{2} < \frac{3 - m^2}{3 + m^2} \leq 1$$

$$\Rightarrow \begin{cases} \dfrac{3 - m^2}{3 + m^2} \leq 1 \Rightarrow 3 - m^2 \leq 3 + m^2 \Rightarrow m^2 \geq 0 \Rightarrow m \in \mathbb{R} \\[4mm] \dfrac{1}{2} < \dfrac{3 - m^2}{3 + m^2} \Rightarrow 3 + m^2 < 6 - 2m^2 \Rightarrow 3m^2 < 3 \Rightarrow m^2 < 1 \Rightarrow -1 < m < 1 \Rightarrow |m| < 1 \end{cases} \stackrel{\cap}{\Rightarrow} |m| < 1$$

Choice (3) is the answer.

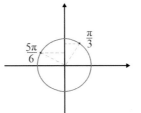

Figure 12.10 The graph of the solution of problem 12.22

12.23. From trigonometry, we know that:

$$\cot(\alpha) = \frac{\cos(\alpha)}{\sin(\alpha)} \tag{1}$$

Based on the information given in the problem, we know that:

$$\sin(\alpha)\cos(\alpha) > 0 \tag{2}$$

$$\cos(\alpha)\cot(\alpha) < 0 \tag{3}$$

Solving (1) and (3):

$$\cos(\alpha)\frac{\cos(\alpha)}{\sin(\alpha)} < 0 \Rightarrow \frac{\cos^2(\alpha)}{\sin(\alpha)} < 0 \xrightarrow{\cos^2(\alpha) > 0} \frac{1}{\sin(\alpha)} < 0 \Rightarrow \sin(\alpha) < 0 \qquad (4)$$

Solving (2) and (4) and considering Figure 12.11:

$$\cos(\alpha) < 0 \qquad (5)$$

Therefore, based on (4) and (5), the end of arc α is in the third quadrant. Choice (3) is the answer.

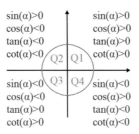

Figure 12.11   The graph of the solution of problem 12.23

12.24. From trigonometry, we know that for an acute angle of θ:

$$\sin(180° - \theta) = \sin(\theta)$$

$$\sin(180° + \theta) = -\sin(\theta)$$

$$\cos(180° - \theta) = -\cos(\theta)$$

$$\cos(180° + \theta) = -\cos(\theta)$$

Moreover, we know that for an acute angle of θ, $\sin(\theta)$ and $\cos(\theta)$ have ascending and descending trends, respectively. Therefore:

Choice (1):

$$\sin(20°) > \sin(170°) \Rightarrow \sin(20°) > \sin(180° - 10°) \Rightarrow \sin(20°) > \sin(10°) \Rightarrow \text{Correct}$$

Choice (2):

$$\cos(20°) < \cos(160°) \Rightarrow \cos(20°) < \cos(180° - 20°) \Rightarrow \cos(20°) < -\cos(20°) \Rightarrow 2\cos(20°) < 0 \Rightarrow \cos(20°) < 0 \Rightarrow \text{Wrong}$$

Choice (3):

$$\sin(20°) < \sin(10°) \Rightarrow \text{Wrong}$$

Choice (4):

$$\cos(10°) < \cos(20°) \Rightarrow \text{Wrong}$$

Choice (1) is the answer.

Reference

1. Rahmani-Andebili, M. (2021). Precalculus – Practice Problems, Methods, and Solutions, Springer Nature, 2021.

Abstract

In this chapter, the basic and advanced problems of arithmetic and geometric sequences are presented. To help students study the chapter in the most efficient way, the problems are categorized into different levels based on their difficulty (easy, normal, and hard) and calculation amounts (small, normal, and large). Moreover, the problems are ordered from the easiest, with the smallest computations, to the most difficult, with the largest calculations.

13.1. If the 12th term and the common difference of successive terms of an arithmetic sequence are 34 and 2, respectively, determine its 18th term [1].

Difficulty level   ● Easy   ○ Normal   ○ Hard

Calculation amount   ● Small   ○ Normal   ○ Large

1) 46
2) 48
3) 50
4) 52

Exercise: Calculate the 10th term of an arithmetic sequence, if the fifth term and the common difference of successive terms of an arithmetic sequence are 20 and 3, respectively.

1) 38
2) 28
3) 35
4) 25

Final answer: Choice (3).

13.2. In an arithmetic sequence, the first term and the common difference of successive terms are 2 and 4, respectively. Determine the ratio of the 11th term to the first term in this arithmetic sequence.

Difficulty level   ● Easy   ○ Normal   ○ Hard

Calculation amount   ● Small   ○ Normal   ○ Large

1) 19
2) 20
3) 21
4) 42

13.3. In the arithmetic sequence of $2, \frac{7}{3}, \ldots$, which term is equal to 6?

Difficulty level   ● Easy   ○ Normal   ○ Hard
Calculation amount   ● Small   ○ Normal   ○ Large
1) 10
2) 11
3) 12
4) 13

Exercise: In the arithmetic sequence of $1, \frac{4}{3}, \ldots$, which term is equal to 9?
1) 23
2) 24
3) 25
4) 26

Final answer: Choice (3).

13.4. The common term of a geometric sequence is presented in the following. Determine the common ratio of successive terms of this geometric sequence.

$$\frac{2}{3 \times 2^n}$$

Difficulty level   ● Easy   ○ Normal   ○ Hard
Calculation amount   ● Small   ○ Normal   ○ Large
1) $\frac{1}{6}$
2) $\frac{1}{3}$
3) $\frac{1}{2}$
4) $\frac{2}{3}$

Exercise: The common term of a geometric sequence is $a_n = \frac{2^n}{5}$. Determine the common ratio of successive terms of this geometric sequence.
1) 1
2) 2
3) 4
4) 8

Final answer: Choice (2).

13.5. In an arithmetic sequence, the first term and the common difference of successive terms are 0.5 and -1, respectively. Calculate the sum of first eight terms of this arithmetic sequence.

Difficulty level   ● Easy   ○ Normal   ○ Hard
Calculation amount   ● Small   ○ Normal   ○ Large
1) -16
2) -20
3) -24
4) -25

> **Exercise:** Calculate the sum of first 10 terms of an arithmetic sequence given that its first term and common difference of successive terms are 3 and 0.5, respectively.
> 1) 55
> 2) 7.5
> 3) 52.5
> 4) 12.5
>
> *Final answer:* Choice (3).

13.6. Which term of the arithmetic sequence 2, 5, 8, ... is equal to 56?

Difficulty level   ● Easy   ○ Normal   ○ Hard
Calculation amount   ● Small   ○ Normal   ○ Large
1) 18
2) 19
3) 20
4) 21

> **Exercise:** Which term of the arithmetic sequence $-12, -9, -6, \ldots$ is equal to 27?
>
> 1) 13
> 2) 14
> 3) 15
> 4) 16
>
> *Final answer:* Choice (2).

13.7. In an arithmetic sequence, $a_1 = 4$ and $a_{n+1} = a_n + 3$. Determine the n-th term of the sequence.

Difficulty level   ● Easy   ○ Normal   ○ Hard
Calculation amount   ● Small   ○ Normal   ○ Large
1) $n + 5$
2) $3n + 1$
3) $2n + 3$
4) $4n - 1$

> **Exercise:** In an arithmetic sequence, $a_1 = -2$ and $a_{n+1} = a_n - 3$. Determine the n-th term of the sequence.
> 1) $-3n - 1$
> 2) $3n - 1$
> 3) $3n + 1$
> 4) $-3n + 1$
>
> *Final answer:* Choice (4).

13.8. In an arithmetic sequence, the sum of the first and the third terms is equal to 1.5 times the sum of the second and the fourth terms. What relation exists between the first term and the common difference of successive terms?

Difficulty level   ● Easy   ○ Normal   ○ Hard
Calculation amount   ● Small   ○ Normal   ○ Large

1) $a_1 = -4d$
2) $d = -4a_1$
3) $d = a_1$
4) $a_1 = -d$

Exercise: In an arithmetic sequence, the first term is 2 and the sum of the third and the sixth terms is equal to two times of the sum of the second and the fifth terms. Calculate the value of common difference of successive terms.

1) $-\dfrac{4}{3}$

2) $\dfrac{4}{3}$

3) $\dfrac{2}{3}$

4) $-\dfrac{2}{3}$

Final answer: Choice (1).

13.9. Calculate the sum of the infinite number of terms of the geometric sequence $12, 8, \frac{16}{3}, \ldots$

Difficulty level   ● Easy   ○ Normal   ○ Hard
Calculation amount   ● Small   ○ Normal   ○ Large

1) 32
2) 36
3) 48
4) 54

Exercise: Calculate the sum of the infinite number of terms of the geometric sequence $-2, 1, -0.5, 0.25, \ldots$

1) -4
2) 4

3) $\dfrac{4}{3}$

4) $-\dfrac{4}{3}$

Final answer: Choice (4).

13.10. Calculate the geometric mean of $\sqrt{15}$ and $\sqrt{240}$.

Difficulty level   ● Easy   ○ Normal   ○ Hard
Calculation amount   ● Small   ○ Normal   ○ Large

1) $3\sqrt{13}$
2) $2\sqrt{13}$
3) $3\sqrt{15}$
4) $2\sqrt{15}$

Exercise: Calculate the geometric mean of 5 and 125.

1) 65
2) 25
3) 5
4) $\sqrt{5}$

Final answer: Choice (2).

13.11. Determine the fifth term of the geometric sequence $4, -6, 9, \ldots$.

Difficulty level   ● Easy   ○ Normal   ○ Hard
Calculation amount   ● Small   ○ Normal   ○ Large

1) $-\dfrac{71}{4}$

2) $-\dfrac{51}{4}$

3) $\dfrac{81}{4}$

4) $\dfrac{61}{4}$

Exercise: Determine the sixth term of the geometric sequence $4, -8, 16, \ldots$.

1) 128
2) -128
3) 256
4) -256

Final answer: Choice (2).

13.12. Calculate the sum of the first six terms of the geometric sequence $32, 16, \ldots$

Difficulty level   ● Easy   ○ Normal   ○ Hard
Calculation amount   ● Small   ○ Normal   ○ Large

1) 58
2) 62
3) 63
4) 66

Exercise: Calculate the sum of the first five terms of the geometric sequence $18, 9, \ldots$

1) $\dfrac{279}{16}$

2) 18

3) 36

4) $\dfrac{279}{8}$

Final answer: Choice (4).

13.13. Calculate the limiting sum of the infinite geometric sequence $-12, 4, -\frac{4}{3}, \ldots$

Difficulty level   ● Easy   ○ Normal   ○ Hard
Calculation amount   ● Small   ○ Normal   ○ Large

1) -3
2) -6
3) -9
4) -18

Exercise: Calculate the sum of the infinite number of terms of the geometric sequence $2, 1, 0.5, \ldots$

1) -4
2) 4
3) $\frac{4}{3}$
4) $-\frac{4}{3}$

Final answer: Choice (2).

13.14. Calculate the geometric mean of $\sqrt{3}$ and $\frac{\sqrt{3}}{4}$.

Difficulty level   ● Easy   ○ Normal   ○ Hard
Calculation amount   ● Small   ○ Normal   ○ Large

1) $\frac{\sqrt{3}}{2}$
2) $\frac{3}{2}$
3) $\frac{\sqrt{3}}{4}$
4) $\frac{3}{4}$

Exercise: Calculate the geometric mean of 2 and 14.

1) $\sqrt{7}$
2) $\sqrt{16}$
3) $2\sqrt{7}$
4) $\sqrt{12}$

Final answer: Choice (3).

13.15. Calculate the arithmetic mean of 3 and 5.

Difficulty level   ● Easy   ○ Normal   ○ Hard
Calculation amount   ● Small   ○ Normal   ○ Large

1) 4
2) 3
3) 5
4) $\sqrt{15}$

> **Exercise:** Calculate the arithmetic mean of 2 and 14.
>
> 1) 16
> 2) 8
> 3) 12
> 4) 6
>
> *Final answer*: Choice (2).

13.16. If the second and the fifth terms of a geometric sequence are 6 and $\frac{16}{9}$, respectively, determine the common ratio of successive terms of the geometric sequence.

Difficulty level   ● Easy   ○ Normal   ○ Hard
Calculation amount   ○ Small   ● Normal   ○ Large

1) $\frac{2}{3}$

2) $\frac{3}{4}$

3) $\frac{4}{3}$

4) $\frac{3}{2}$

> **Exercise:** Calculate the common ratio of successive terms of a geometric sequence when its third and seventh terms are 6 and 9, respectively.
>
> 1) $\sqrt[4]{\frac{3}{2}}$
>
> 2) $\sqrt{\frac{3}{2}}$
>
> 3) $\sqrt[4]{\frac{2}{3}}$
>
> 4) $\sqrt{\frac{2}{3}}$
>
> *Final answer*: Choice (1).

13.17. Determine the first term of an arithmetic sequence if the sum of its first 12 terms is 120 and the common difference of successive terms is 2.

Difficulty level   ○ Easy   ● Normal   ○ Hard
Calculation amount   ● Small   ○ Normal   ○ Large

1) −2
2) −1
3) 1
4) 2

Exercise: Calculate the common difference of successive terms of an arithmetic sequence when the sum of its first 10 terms is 100 and the first term is 3.

1) $\dfrac{196}{9}$

2) 20

3) $\dfrac{7}{5}$

4) $\dfrac{14}{9}$

Final answer: Choice (4).

13.18. In a geometric sequence, the sum of the first and third terms is equal to 1.5 times the sum of the second and fourth terms. Determine the common ratio of the successive terms.

Difficulty level   ○ Easy   ● Normal   ○ Hard
Calculation amount   ● Small   ○ Normal   ○ Large

1) $\dfrac{1}{3}$

2) $\dfrac{1}{2}$

3) $\dfrac{2}{3}$

4) $\dfrac{3}{2}$

13.19. In a geometric sequence, the subtraction of the fifth and third terms is equal to $\frac{1}{32}$. If the common ratio of successive terms is $\frac{1}{2}$, determine the first term of the geometric sequence.

Difficulty level   ○ Easy   ● Normal   ○ Hard
Calculation amount   ● Small   ○ Normal   ○ Large

1) $-\dfrac{2}{3}$

2) $-\dfrac{1}{6}$

3) $\dfrac{1}{8}$

4) $\dfrac{3}{2}$

13.20. Calculate the common ratio of successive terms of a descending geometric sequence given that its limiting sum is equal to four times the sum of the first two terms.

Difficulty level   ○ Easy   ● Normal   ○ Hard
Calculation amount   ● Small   ○ Normal   ○ Large

1) $\dfrac{\sqrt{3}}{4}$

2) $\dfrac{1}{4}$

3) $\dfrac{\sqrt{3}}{2}$

4) $\dfrac{3}{4}$

Exercise: Calculate the common ratio of successive terms of a descending geometric sequence given that its limiting sum is equal to three times the sum of the first two terms.

1) $\sqrt{\frac{2}{3}}$

2) $\sqrt{\frac{3}{2}}$

3) $\sqrt{\frac{4}{3}}$

4) $\sqrt{\frac{3}{4}}$

Final answer: Choice (1).

13.21. In a geometric sequence, the first and seventh terms are 10 and 640, respectively. Calculate the sum of the first six terms of the sequence.

Difficulty level   ○ Easy   ● Normal   ○ Hard
Calculation amount   ○ Small   ● Normal   ○ Large

1) 1260
2) 640
3) 730
4) 630

13.22. Which one of the following successive terms belongs to an arithmetic sequence?

Difficulty level   ○ Easy   ● Normal   ○ Hard
Calculation amount   ○ Small   ● Normal   ○ Large

1) $\frac{1}{\sqrt{5}}, 1, \sqrt{5}$

2) $\frac{5}{3}, 1, \frac{2}{3}$

3) $\frac{9}{5}, \frac{6}{5}, \frac{3}{5}$

4) $\frac{27}{5}, \frac{9}{5}, \frac{3}{5}$

Exercise: Which one of the following successive terms belongs to an arithmetic sequence?

1) $\frac{19}{3}, \frac{19}{2}, 19$

2) $-1, 0, 1$

3) $\frac{1}{\sqrt{2}}, 1, \sqrt{2}$

4) $1, 2, 5$

Final answer: Choice (2).

Exercise: Which one of the following successive terms belongs to a geometric sequence?

1) $\dfrac{19}{3}, \dfrac{19}{2}, 19$

2) $-1, 0, 1$

3) $\dfrac{1}{\sqrt{2}}, 1, \sqrt{2}$

4) $1, 2, 5$

Final answer: Choice (3).

13.23. Calculate the common difference of successive terms of an arithmetic sequence when its 7th and 15th terms are 20 and 30, respectively.

Difficulty level   ○ Easy   ● Normal   ○ Hard
Calculation amount   ○ Small   ● Normal   ○ Large

1) $\dfrac{3}{5}$

2) $\dfrac{4}{5}$

3) $\dfrac{5}{4}$

4) $\dfrac{5}{3}$

13.24. The seventh and ninth terms of an arithmetic sequence are 15 and 19, respectively. Calculate the 12th term of this arithmetic sequence.

Difficulty level   ○ Easy   ● Normal   ○ Hard
Calculation amount   ○ Small   ● Normal   ○ Large

1) 24
2) 25
3) 27
4) 29

Exercise: The fifth and eighth terms of an arithmetic sequence are 10 and 4, respectively. Determine the 10th term of this arithmetic sequence.

1) 0
2) 1
3) 2
4) 3

Final answer: Choice (1).

13.25. The terms $\frac{1}{a}, \frac{1}{b}, and \frac{1}{c}$ are the successive terms of a geometric sequence. Which one of the following statements is true about the successive terms of $\log a$, $\log b$, and $\log c$?

Difficulty level   ○ Easy   ○ Normal   ● Hard
Calculation amount   ● Small   ○ Normal   ○ Large

1) They are the successive terms of an arithmetic sequence
2) They are the successive terms of a geometric sequence
3) They can be the successive terms of an arithmetic or a geometric sequence
4) They are not the successive terms of a sequence

Reference

1. Rahmani-Andebili, M. (2021). Precalculus – Practice Problems, Methods, and Solutions, Springer Nature, 2021.

Abstract

In this chapter, the problems of the 13th chapter are fully solved, in detail, step-by-step, and with different methods.

14.1. As we know, the common term of an arithmetic sequence is as shown below [1]:

$$a_n = a_1 + (n-1)d \qquad (1)$$

Based on the information given in the problem, we have:

$$a_{12} = 34 \qquad (2)$$

$$d = 2 \qquad (3)$$

Solving (1), (2), and (3):

$$34 = a_1 + (12-1) \times 2 \Rightarrow a_1 = 12 \qquad (4)$$

Solving (1), (3), and (4):

$$a_{18} = 12 + (18-1) \times 2 = 46$$

Choice (1) is the answer.

14.2. As we know, the common term of an arithmetic sequence is as shown below:

$$a_n = a_1 + (n-1)d$$

Based on the information given in the problem, we have:

$$a_1 = 2$$

$$d = 4$$

M. Rahmani-Andebili, *Precalculus*, https://doi.org/10.1007/978-3-031-49364-5_14

Therefore:

$$\frac{a_{11}}{a_1} = \frac{a_1 + (n-1)d}{a_1} = \frac{2 + (11-1) \times 4}{2} = \frac{42}{2} = 21$$

Choice (3) is the answer.

14.3. From the arithmetic sequence $2, \frac{7}{3}, \ldots$, the following information can be extracted.

$$a_1 = 2 \qquad (1)$$

$$d = \frac{7}{3} - 2 = \frac{1}{3} \qquad (2)$$

Moreover, the information below is given in the problem.

$$a_n = 6 \qquad (3)$$

As we know, the common term of an arithmetic sequence is as shown below:

$$a_n = a_1 + (n-1)d \qquad (4)$$

Solving (1)–(4):

$$6 = 2 + (n-1) \times \frac{1}{3} \Rightarrow (n-1) \times \frac{1}{3} = 4 \Rightarrow n - 1 = 12 \Rightarrow n = 13$$

Choice (4) is the answer.

14.4. Based on the information given in the problem, we have:

$$a_n = \frac{2}{3 \times 2^n}$$

As we know, the common ratio of successive terms of a geometric sequence can be determined as follows:

$$q = \frac{a_{n+1}}{a_n}$$

Therefore:

$$q = \frac{a_2}{a_1} = \frac{\frac{2}{3 \times 2^2}}{\frac{2}{3 \times 2^1}} = \frac{\frac{1}{3 \times 2}}{\frac{1}{3}} = \frac{1}{2}$$

Choice (3) is the answer.

14.5. Based on the information given in the problem, we have:

$$a_1 = \frac{1}{2}$$

$$d = -1$$

As we know, the sum of the first n terms of an arithmetic sequence can be calculated as follows:

$$S_n = \frac{n}{2}(2a_1 + (n-1)d)$$

Therefore:

$$S_8 = \frac{8}{2}\left(2 \times \frac{1}{2} + (8-1) \times (-1)\right) = 4(1-7) = -24$$

Choice (3) is the answer.

14.6. As we know, the common term of an arithmetic sequence is as shown below:

$$a_n = a_1 + (n-1)d \tag{1}$$

By looking at the arithmetic sequence 2, 5, 8, ..., the following information is achieved.

$$a_1 = 2 \tag{2}$$

$$d = 5 - 2 = 3 \tag{3}$$

Solving (1), (2), and (3) for $a_n = 56$:

$$56 = 2 + (n-1) \times 3 \Rightarrow n - 1 = \frac{54}{3} = 18 \Rightarrow n = 19$$

Choice (2) is the answer.

14.7. As we know, the common term of an arithmetic sequence is as shown below:

$$a_n = a_1 + (n-1)d \tag{1}$$

Based on the information given in the problem, we have:

$$a_1 = 4 \tag{2}$$

$$a_{n+1} = a_n + 3 \tag{3}$$

The common difference of successive terms can be calculated as follows:

$$d = a_{n+1} - a_n = 3 \tag{4}$$

By solving (1), (2), and (4), the n-th term of the sequence can be determined.

$$a_n = 4 + (n-1) \times 3 = 3n + 1$$

Choice (2) is the answer.

14.8. Based on the information given in the problem, we have:

$$a_1 + a_3 = 1.5(a_2 + a_4) \tag{1}$$

As we know, the common term of an arithmetic sequence is as shown below:

$$a_n = a_1 + (n-1)d \tag{2}$$

Solving (1) and (2):

$$a_1 + a_1 + (3-1)d = 1.5(a_1 + (2-1)d + a_1 + (4-1)d) \Rightarrow 2a_1 + 2d = 1.5(2a_1 + 4d)$$

$$\Rightarrow a_1 = -4d$$

Choice (1) is the answer.

14.9. By looking at the geometric sequence $12, 8, \frac{16}{3}, \ldots$, the following results can be achieved.

$$a_1 = 12$$

$$q = \frac{8}{12} = \frac{2}{3}$$

The sum of infinite number of terms of a geometric sequence, where the magnitude of its common ratio of successive terms is less than 1, can be calculated as follows:

$$\lim_{n \to \infty} S_n = \frac{a_1}{1-q}, \quad |q| < 1$$

Therefore:

$$\lim_{n \to \infty} S_n = \frac{a_1}{1-q} = \frac{12}{1-\frac{2}{3}} = 36$$

Choice (2) is the answer.

14.10. In a geometric sequence, the geometric mean ($G.~M.$) of a and b can be calculated as follows:

$$G.M. = \sqrt{a \times b}$$

Therefore:

$$G.M. = \sqrt{\sqrt{15} \times \sqrt{240}} = \sqrt{\sqrt{15} \times \sqrt{16 \times 15}} = \sqrt{\sqrt{15} \times 4\sqrt{15}} = \sqrt{4 \times 15} = 2\sqrt{15}$$

Choice (4) is the answer.

14.11. As we know, the common term of a geometric sequence is as shown below:

$$a_n = a_1 q^{n-1} \qquad (1)$$

Moreover, the common ratio of successive terms of a geometric sequence can be determined as follows:

$$q = \frac{a_{n+1}}{a_n} \qquad (2)$$

From the geometric sequence $4, -6, 9, \ldots$, the information below can be extracted.

$$a_1 = 4 \qquad (3)$$

$$q = \frac{a_{n+1}}{a_n} = \frac{-6}{4} = -\frac{3}{2} \qquad (4)$$

Solving (1), (3), and (4):

$$a_5 = 4\left(-\frac{3}{2}\right)^{5-1} = 4 \times \left(\frac{81}{16}\right) = \frac{81}{4}$$

Choice (3) is the answer.

14.12. From the geometric sequence $32, 16, \ldots$, the following results can be concluded.

$$a_1 = 32$$

$$q = \frac{16}{32} = \frac{1}{2}$$

The sum of finite number of terms of a geometric sequence can be calculated as follows:

$$S_n = \frac{a_1(1 - q^n)}{1 - q}$$

Therefore:

$$S_6 = \frac{32\left(1 - \left(\frac{1}{2}\right)^6\right)}{1 - \frac{1}{2}} = \frac{32\left(1 - \frac{1}{64}\right)}{\frac{1}{2}} = \frac{32 \times \frac{63}{64}}{\frac{1}{2}} = 63$$

Choice (3) is the answer.

14.13. By looking at the geometric sequence $-12, 4, -\frac{4}{3}, \ldots$, the following results are achieved.

$$a_1 = -12$$

$$q = \frac{4}{-12} = -\frac{1}{3}$$

The limiting sum of the infinite geometric sequence can be calculated as follows:

$$S = \frac{a_1}{1-q}, \quad |q| < 1$$

Therefore:

$$S = \frac{a_1}{1-q} = \frac{-12}{1-\left(-\frac{1}{3}\right)} = \frac{-12}{\frac{4}{3}} = -9$$

Choice (3) is the answer.

14.14. In a geometric sequence, the geometric mean (*G. M.*) of *a* and *b* can be calculated as follows:

$$G.M. = \sqrt{a \times b}$$

Therefore:

$$G.M. = \sqrt{\sqrt{3} \times \frac{\sqrt{3}}{4}} = \sqrt{\frac{3}{4}} = \frac{\sqrt{3}}{2}$$

Choice (1) is the answer.

14.15. In an arithmetic sequence, the arithmetic mean (*A. M.*) of *a* and *b* can be calculated as follows:

$$A.M. = \frac{a+b}{2}$$

Therefore:

$$A.M. = \frac{a+b}{2} = \frac{3+5}{2} = 4$$

Choice (1) is the answer.

14.16. Based on the information given in the problem, we have:

$$a_2 = 6 \tag{1}$$

$$a_5 = \frac{16}{9} \tag{2}$$

As we know, the common term of a geometric sequence is as shown below:

$$a_n = a_1 q^{n-1} \tag{3}$$

Solving (1) and (3):

$$6 = a_1 q^{2-1} \Rightarrow 6 = a_1 q \tag{4}$$

Solving (2) and (3):

$$\frac{16}{9} = a_1 q^{5-1} \Rightarrow \frac{16}{9} = a_1 q^4 \tag{5}$$

Solving (4) and (5):

$$\frac{a_1 q^4}{a_1 q} = \frac{\frac{16}{9}}{6} \Rightarrow q^3 = \frac{8}{27} \Rightarrow q = \frac{2}{3}$$

Choice (1) is the answer.

14.17. As we know, the sum of the first n terms of an arithmetic sequence can be calculated using the formula below.

$$S_n = \frac{n}{2}(2a_1 + (n-1)d)$$

Based on the information given in the problem, we know that:

$$n = 12$$

$$S_{12} = 120$$

$$d = 2$$

Therefore:

$$120 = \frac{12}{2}(2a_1 + (12-1) \times 2) \Rightarrow 20 = 2a_1 + 22 \Rightarrow -2 = 2a_1 \Rightarrow a_1 = -1$$

Choice (2) is the answer.

14.18. Based on the information given in the problem, we have:

$$a_1 + a_3 = 1.5(a_2 + a_4) \tag{1}$$

As we know, the common term of a geometric sequence is as shown below:

$$a_n = a_1 q^{n-1} \tag{2}$$

Solving (1) and (2):

$$a_1 + a_1 q^{3-1} = 1.5(a_1 q^{2-1} + a_1 q^{4-1}) \Rightarrow 1 + q^2 = 1.5q(1 + q^2) \Rightarrow 1 = 1.5q \Rightarrow q = \frac{2}{3}$$

Choice (3) is the answer.

14.19. Based on the information given in the problem, we have:

$$a_5 - a_3 = \frac{1}{32} \tag{1}$$

$$q = \frac{1}{2} \tag{2}$$

As we know, the common term of a geometric sequence is as shown below:

$$a_n = a_1 q^{n-1} \tag{3}$$

Solving Eqs. (1)–(3):

$$a_1 \left(\frac{1}{2}\right)^{5-1} - a_1 \left(\frac{1}{2}\right)^{3-1} = \frac{1}{32} \Rightarrow a_1 \left(\left(\frac{1}{2}\right)^4 - \left(\frac{1}{2}\right)^2 \right) = \frac{1}{32} \Rightarrow a_1 = \frac{\frac{1}{32}}{\frac{1}{16} - \frac{1}{4}} = \frac{\frac{1}{32}}{-\frac{3}{16}} = -\frac{1}{6}$$

Choice (2) is the answer.

14.20. Based on the information given in the problem, we have:

$$S_\infty = 4(a_1 + a_2) \tag{1}$$

The common term of a geometric sequence is as shown below:

$$a_n = a_1 q^{n-1} \tag{2}$$

In addition, the limiting sum of the infinite geometric sequence can be calculated as follows:

$$S = \frac{a_1}{1-q}, \quad |q| < 1 \tag{3}$$

Therefore:

$$\frac{a_1}{1-q} = 4(a_1 + a_1 q) \Rightarrow \frac{1}{1-q} = 4(1+q) \Rightarrow 1 - q^2 = \frac{1}{4} \Rightarrow q^2 = \frac{3}{4} \Rightarrow q = \frac{\sqrt{3}}{2}$$

Choice (3) is the answer.

14.21. Based on the information given in the problem, we have:

$$a_1 = 10 \tag{1}$$

$$a_7 = 640 \tag{2}$$

Moreover, the common term of a geometric sequence is as shown below:

$$a_n = a_1 q^{n-1} \tag{3}$$

The sum of finite number of terms of a geometric sequence can be calculated as follows:

$$S_n = \frac{a_1(1-q^n)}{1-q} \tag{4}$$

Solving (1), (2), and (3):

$$640 = 10q^{7-1} \Rightarrow q^6 = 64 \Rightarrow q = 2 \tag{5}$$

Solving (1), (4), and (5):

$$S_6 = \frac{10(1-2^6)}{1-2} = \frac{10(-63)}{-1} = 630$$

Choice (4) is the answer.

14.22. If three successive terms belong to an arithmetic sequence, the relation below, which is used to calculate the arithmetic mean of the successive terms of a_1 and a_3, must be held.

$$a_2 = \frac{a_1 + a_3}{2}$$

Choice (1):

$$1 \neq \frac{\frac{1}{\sqrt{5}} + \sqrt{5}}{2} \Rightarrow 1 \neq \frac{3\sqrt{5}}{5}$$

Choice (2):

$$1 \neq \frac{\frac{5}{3} + \frac{2}{3}}{2} \Rightarrow 1 \neq \frac{7}{6}$$

Choice (3):

$$\frac{6}{5} = \frac{\frac{9}{5} + \frac{3}{5}}{2} \Rightarrow \frac{6}{5} = \frac{6}{5}$$

Choice (4):

$$\frac{9}{5} \neq \frac{\frac{27}{5} + \frac{3}{5}}{2} \Rightarrow \frac{9}{5} \neq 3$$

Choice (3) is the answer.

14.23. As we know, the common term of an arithmetic sequence is as shown below:

$$a_n = a_1 + (n-1)d \tag{1}$$

Based on the information given in the problem, we have:

$$a_7 = 20 \tag{2}$$

$$a_{15} = 30 \tag{3}$$

Solving (1), (2), and (3):

$$\begin{cases} 20 = a_1 + (7-1)d \\ 30 = a_1 + (15-1)d \end{cases} \Rightarrow \begin{cases} 20 = a_1 + 6d \\ 30 = a_1 + 14d \end{cases} \Rightarrow 10 = 8d \Rightarrow d = \frac{5}{4}$$

Choice (3) is the answer.

14.24. Based on the information given in the problem, we have:

$$a_7 = 15 \tag{1}$$

$$a_9 = 19 \tag{2}$$

Moreover, as we know, the common term of an arithmetic sequence is as shown below:

$$a_n = a_1 + (n-1)d \tag{3}$$

Therefore:

$$\begin{cases} 15 = a_1 + (7-1)d \\ 19 = a_1 + (9-1)d \end{cases} \Rightarrow \begin{cases} 15 = a_1 + 6d \\ 19 = a_1 + 8d \end{cases} \overset{-}{\Rightarrow} 4 = 2d \Rightarrow d = 2 \tag{4}$$

$$15 = a_1 + 6d \overset{d=2}{\Longrightarrow} 15 = a_1 + 12 \Rightarrow a_1 = 3 \tag{5}$$

Solving (3), (4), and (5) for $n = 12$:

$$a_{12} = 3 + (12-1) \times 2 = 3 + 22 = 25$$

Choice (2) is the answer.

14.25. In a geometric sequence, the relation below is held, where a_2 is the geometric mean of the successive terms of a_1, a_2, and a_3 if:

$$(a_2)^2 = a_1 a_3 \tag{1}$$

Moreover, in an arithmetic sequence, the relation below is held, where a_2 is the arithmetic mean of the successive terms of a_1, a_2, and a_3 if:

$$a_2 = \frac{a_1 + a_2}{2} \tag{2}$$

In addition, we know that the relations below are held for logarithm.

$$\log(x^2) = 2\log(x) \tag{3}$$

$$\log(xy) = \log(x) + \log(y) \tag{4}$$

Based on the information given in the problem, we know that the terms $\frac{1}{a}$, $\frac{1}{b}$, and $\frac{1}{c}$ are the successive terms of a geometric sequence. Therefore, by considering (1), we can write:

$$\left(\frac{1}{b}\right)^2 = \frac{1}{a} \times \frac{1}{c} \Rightarrow b^2 = ac \tag{5}$$

$$\overset{Log}{\Longrightarrow} \log(b^2) = \log(ac) \Rightarrow 2\log(b) = \log(a) + \log(c) \Rightarrow \log(b) = \frac{\log(a) + \log(c)}{2}$$

Therefore, by considering (2), we see that the successive terms of $\log a$, $\log b$, and $\log c$ belong to an arithmetic sequence. Choice (1) is the answer.

Reference

1. Rahmani-Andebili, M. (2021). Precalculus – Practice Problems, Methods, and Solutions, Springer Nature, 2021.

Index